Other Books by Jerry Pollock

Divinely Inspired: Spiritual Awakening of a Soul

Messiah Interviews: Belonging to God

Gog and Magog: The Devil's Descendants

The Wing of the Butterfly: Memories of Marcia

I0167211

REVIEWS FOR
PUTTING GOD INTO EINSTEIN'S EQUATIONS

"What makes *Putting God Into Einstein's Equations: Energy of the Soul* by Jerry and Marcia Pollock, so compelling is that the reader can discern that they genuinely believe what they say and have the courage of their convictions. Their writing reflects penetrating, sensible thoughts, and their words are not the random ramblings of maniacal madmen. Truth, love, and wisdom lie within these pages, and the moving reflections on the authors' life experiences will either leave you entertained or enlightened, depending on your frame of mind at Page 1. Either way, you will be enthralled by the end of the book because you will have confronted dozens of thought-provoking theories, and weighing the pros and cons of each conjecture may keep you up half the night. This book is a page turner, and most people will desire to read the book in one mind-expanding sitting. In reading this book, you set sail on a spellbinding trek to a greater understanding of who we are, why we are here, and where we are heading. At times, the book reaches sublime proportions. Prepare yourself for an unprecedented excursion skyward."

***** *David Menafee, Book Pleasures*

"After discovering the premise for the Jerry and Marcia Pollock's book, *Putting God Into Einstein's Equations: Energy of the Soul*, I was prepared to read it with an open mind, as my religious and philosophical views somewhat differ from the content held within these pages. However, what I found at the essence of the book was a story that I could relate to, a tale of true romance and of a strong faith and devotion to our God and to each other. What are the properties of the soul? What happens after we die? Can we be once again be reunited with those beyond us? These are just some of the intricate concepts that Pollock explores using Einstein's equations as a platform to bring us an understanding of God, our souls, our universe, and ourselves. I found that Marcia as a coauthor from the spirit world to be a unique premise upon which the book is written. The relationship that he had with her in both the physical and spiritual world is a beautiful thing; often books do not portray the emotions as well as this book did. It is a true tale of how faith and love can bring us to a real understanding, in this life and the next."

***** *Susan Gattis, Pacific Book Review*

"The Pollocks unabashedly delve into the realm of attempting to definitively explain and even quantify God, Satan, heaven, reincarnation, biblical prophecies, and more. Further, they both claim to have been blessed by personal conversations with God and feel that they, their family, and some of their close friends are integral to the imminence of the Messianic Age. Some of the book's premises are so extraordinary that one has to set aside judgment and take a breath before reading on, possibly separating the useful and meaningful from the bizarre and untenable. *Putting God into Einstein's Equations: Energy of the Soul* is also a stunningly complex extrapolation of Albert Einstein's theories and spiritual musings, and lovers of Einstein's mystical side would find this an excellent read."

**** *Peggy Sutherland, ForeWord Clarion Reviews*

"Putting God Into Einstein's Equations: Energy of the Soul" is a deep love story mixed in with understanding quantum physics,

insight into Albert Einstein, and the meaning of life. Taking the reader on a journey that clearly shows there is no space and time when it comes to love, Jerry shares intimate details that show absolutely why their paths crossed in this life time. A deep intriguing read, this book is perfect for those who refuse to take life at face value. If you ever thought that you would find your soul mate, this book will convince you it is absolutely possible."

***** *Kathleen Gage, Street Smarts Marketing*

"I started with Weiss – Past Lives, advanced to Newton – Spirit World, and finally Jerry Pollock has gone where no one else has dared to travel. In his book, *Putting God Into Einstein's Equations: Energy of the Soul*, Pollock has explored past lives and the spirit world through direct telepathic communication. One must keep an open mind as Pollock is brutally honest about his life which makes him real and honest as he speaks to the reader. Having read books by Weiss and Newton, I have progressed in my thinking whereby I am totally open to what Dr. Pollock has written. Take the next step, keep an open mind and you will be exposed to new concepts and ideas."

***** *Bruce Herman, Book Reader*

"*Putting God Into Einstein's Equations: Energy of the Soul* is a book written by a man who lost his wife too soon in her journey on earth. Jerry Pollock in conjunction with the essence of all that Marcia is, presents a philosophy of eternal life based on their faith and science. Marcia Pollock died March 18, 2011 leaving behind a wealth of wisdom written for all to see. Anyone who has experienced the loss of a loved one, particularly a spouse, understands some of what Jerry experienced. Marcia was the one who made him whole; this is very clear as you journey with him in the pages of the book. Though I cannot recommend this book from a Christian perspective, I want you to go buy it. I don't think you will ever read a greater love story blended in with faith and science."

***** *Marsha Randolph, No Wayz Tired Book Blog*

"I found this book of spiritual investigation and growth intriguing, fascinating and provocative. The collaboration between these soul mates is evidence that true love never dies. I know very little about the theories of Albert Einstein. I did find the thoughts expressed in this book very interesting. Dr. and Mrs. Pollock used examples from research and their own experiences to illustrate and explain their beliefs. The following is one of my favorite passages from the book. It reminds us that spiritual growth takes time and effort; 'In the course of life, humans seem to easily be able to darken their Divine souls. It is much easier to blacken your soul than it is to brighten it. Climbing the spiritual ladder takes time and energy. Living in the material world is impulsive.' This book might not be for everyone, but I recommend it for all who are serious about spiritual growth and development. This book will give you insight and guidance as you climb the spiritual ladder."

***** *Janette Fuller, Books, Inspiration, Life Blog*

"Faith does not have to contend with science. *Putting God Into Einstein's Equations* is an exploration of science and the soul from Jerry and his late wife Marcia as they explore the theories of Einstein from the perspectives of faith, souls, with some metaphysical ideals woven in as well. "Putting God Into Einstein's Equations" is worth considering for those who want to weave it all together into something much more. Highly recommended."

***** *James A. Cox, Midwest Book Review*

PUTTING GOD
into
EINSTEIN'S EQUATIONS

PUTTING GOD
into
EINSTEIN'S
EQUATIONS

ENERGY OF THE SOUL

sHecHinaH

THIRD TEMPLE

JERRY & MARCIA POLLOCK

SHECHINAH THIRD TEMPLE, INC.

Putting God into Einstein's Equations is a unique nonfiction book written by a living Jerry Pollock and his deceased spouse, Marcia Pollock. Husband and wife have communicated across two worlds, the physical and the spiritual, through the energy of their Divine souls.

Published by Shechinah Third Temple, Inc.

ISBN: 0972386661

ISBN, Softbound Edition: 978-09723866-6-1

EAN-13: 9780972386661

Library of Congress Control Number: 2012900927

Publisher's Cataloguing-in Publication

Pollock, Jerry J. (Jerry Joseph), 1941-

Putting God into Einstein's Equations : energy of the soul / Jerry and Marcia Pollock. – Boynton Beach, FL : Shechinah Third Temple, Inc., c2012.

p. ; cm.

ISBN: 978-09723866-6-1

Includes bibliographical references.

1. God–Attributes. 2. Einstein, Albert, 1879-1955. 3. Relativity Physics)–Religious aspects. 4. Telepathy. 5. Soul. 6. Spirituality. 7. Incarnation. 8. Devil–Attributes. I. Pollock, Marcia (Marcia Kay), 1942-2011. II. Title.

BL205.P652012   2012900927
212/.7-dc23   1204

CreateSpace, Charleston, SC

Printed in the United States of America

Walter Isaacson
"Einstein: His Life and Universe"
Einstein on God: *Try and penetrate with a limited
means the secrets of nature and you will find
that behind all the discernible laws and
connections, there remains something subtle,
intangible and inexplicable. We all dance to
a mysterious tone intoned in the distance by an
invisible player* [GOD].

Jerry Pollock
"Divinely Inspired" and "Messiah Interviews"
*Man achieves the possible. God accomplishes
the impossible. Miracles cannot explain science,
nor can science explain miracles; yet, God created
our universe and life through scientific miracles.*

Marcia Pollock
"The Wing of the Butterfly"
27th Wedding Anniversary, 3 months before my death:
Jerry, I shall be with you always and be there forever.

For All Eternity

Marcia, I am your Adam
and you are my Eve,
and we shall dwell only
in the Garden of Eden for
our souls are one with God.

Our Creator

We love You with our
heart, soul, and might.
We honor You by passing on
righteousness and justice to
our children, grandchildren,
and to the generations
that follow.

PUBLISHER'S NOTE

A BIBLICAL COMMENTATOR ONCE WROTE, *"EVEN THE MOST OBJECTIVE PEOPLE HAVE BLIND SPOTS WHERE THEIR OWN INTERESTS ARE CONCERNED."* You the reader have a decision to make based upon your God-given free will. You can read this book with blinders and dismiss its contents, or you can open up your mind to the possibility that telepathic communication is possible between Marcia and me across two worlds, the physical and the spiritual. You will also need to make a leap of faith to believe that this communication is through Divine energy transmission. Our energy is fully measurable within each of and between our two souls, even if one of our souls, Marcia Pollock's, is in the spirit world and no longer in her body.

Einstein believed in an all-powerful God of the universe whose organizing intelligence is beyond human comprehension. He stressed the imagination of the individual as being superior to the knowledge provided by learned men and women over the course of history. He felt that using one's imagination would allow one to solve a problem at a different level than the level at which the problem was created. His theory of relativity has stood the test of time because the theory holds true even when you divorce it from the person, Einstein, who proposed it.

The authors are not physicists like the brilliant Albert Einstein, far from it. However, Jerry Pollock had a thirty-five year career in scientific research, while Marcia Pollock has the luxury of now living in the soul world with a full awareness and knowledge of the physical world with its unique blend of soul and body that allows human function. What we do claim is that we are conveying truth based upon our energy measurements and collaboration of our Divine souls. Einstein defined energy in physical and mathematical terms in his wonderfully innovative theoretical equations. We simply suggest that Einstein's energy equations reflect the energy of our God-given Divine souls and we hope to demonstrate this to the reader. Forgive us if we as authors unintentionally err, but in our book, *Putting God into Einstein's Equations: Energy of the Soul*, we are attempting to join spirituality with science and place God at the forefront of all Creation and Evolution.

For further correspondence, contact Shechinah Third Temple, Inc.

Tel: 561-735-7958
Fax: 561-738-1535
Email: jerrypollock@bellsouth.net
         thirdtemple@bellsouth.net
Web: www.shechinahthirdtemple.org
         www.jerrypollock.com

ACKNOWLEDGMENTS

EVERYTHING MARCIA AND I WERE, ARE, AND WILL EVER BE IN PAST, PRESENT, AND FUTURE LIVES, WE OWE TO OUR CREATOR. For Marcia her belief in God came at an early age through the teachings of her mother. For me my faith came late in life after God's intervention through Divine miracles. For both of us we are grateful to the Creator for bringing us together even in death across two worlds and over the centuries of many past lives. We hope we are giving back to our Holy Creator through our humble attempt to understand His Essence and thereby tell as much as we possibly can about His story in *Putting God into Einstein's Equations.* We thank God for His endorsement on the front cover of our book.

As strange and gory as it sounds, this book could not have been written without the passing of Marcia. Jerry thanks Marcia for her greatest sacrifice, self-sacrifice on my behalf. She ultimately died in 2011, sixteen years after her pledge to God to give her life up for me so that I could emerge from a severe, hopeless bipolar agitated depression that relentlessly persisted for eight months in 1995. With her passing, which has brought on unbearable sadness and grief, I had a yearning to make contact with her soul in the spirit world. I saw a hypnotist by the fictional name of Jules. Through Jules's expertise and gracious help, I twice experienced past life regressions like thousands of other people. When I died in these past life experiences under hypnosis, my soul rose to meet up with Marcia's in the Heavenly World of Souls.

We are grateful for our daughter Melanie's unwavering support and contribution in our good and evil energy measurements. Jerry thanks his former Ph.D. student, Thomas Spradley, for his scientific insights. Jerry also wishes to acknowledge our accountant and friend, Bob McGrath, who recommended reading Walter Isaacson's biography of *Einstein: His Life and Universe*. I read Isaacson's wonderful complete description of Einstein's life after I read Russell Stannard's excellent book, *Relativity: A Very Short Introduction*. There is a saying, *"Without Einstein we would not understand our world. Without God there would be no world to understand."* Einstein's memorable imaginative equations on energy form the basis for the writing of this book. They are the foundations upon which Marcia and I insert God and His spiritual world into Einstein's world of physics and mathematics. In thanking Einstein, we suggest an addendum to the quote above: *"Without Albert Einstein we could not have written this book. Without God there would be no impetus for us to take on such an awesome undertaking."*

We gratefully acknowledge the essential help and support from the CreateSpace Team including Elizabeth my excellent copyeditor and Sonia my amazing cover designer. They are unquestionably fabulous in their serious approach, attitude, expertise, work ethic, input, and guidance, all of which result in the production of books of superior professional quality. We recommend them highly.

1

❧

STUMPED?

M Y MEMOIR, *DIVINELY INSPIRED SPIRITUAL AWAKENING OF A SOUL* WAS PUBLISHED IN 2003. Five years earlier the sparks of my soul had been activated through Divine Providence. I was a novice at computers back then and I enlisted the help of my colleague, Dr. Philias Garant, at Stony Brook University Dental School, who was writing his own book and was familiar with using publishing software to set up interior text.

During this period of my writing, Phil and I would have discussions about the content of my book, which was written as a testament to the presence of God based upon His intervention into my life with Divine miracles. Like many biologists, Phil was skeptical about God. However way before the miracles, as far back as my bar mitzvah, I had a "foggy" faith inside of me. This was true despite that during most of my life I was spiritually bankrupt.

By 2003 I thought that the crux of the spiritual problem was whether you believed or did not believe in God. However, when Phil sent me a zinger of a question, I was stumped. I had vigorously defended the position that my and others' spiritual experiences were evidence for the Creator being in our present world. I argued this was absolutely true despite that God was no longer on the public stage of our earthly life as he was back in biblical days. Phil responded effortlessly with a certain amount of gusto:

And where did he [God] come from?

For almost the last nine years, I have been turning over Phil's question in my mind. Yet I have never had a way to even attempt to answer it. Einstein was right. You have to use your imagination to find solutions, so what better way to draw nearer to an answer than to visit the imaginative world of the spirit, where Marcia's soul was now living after her death? Nine months after her passing, a strengthening connection between Marcia's soul and my soul has been established whereby we now freely telepathically communicate as if she were still alive.

How is this possible?

The world has many looming questions that have never been answered. A trite philosophical one might be, *"What came first the chicken or the egg?"* Or a more scientific query might be, *"Which came first in the origin of evolution, DNA or enzyme proteins?* Like the chicken and the egg conundrum where you need the chicken to hatch the egg and the egg to hatch the chicken, you need enzyme proteins to help make DNA and DNA to help make these same enzyme proteins. The crux of the problem in terms of my friend Phil's question about where did God come from might be:

What existed in the time slot
before the universe was created?
And what was God's involvement
before and after?

2

The first half of the question of "before" has never been satisfactorily answered by scientists, even though the big bang and other unique scientific theories have been proposed to explain the "after." Neither has God's role been deciphered on both ends, although explanations have been offered these past centuries by the mystics of the kabbalah. The only hint we have from the Book of Genesis, as translated by Mesorah Publications, is that on the *First Day,* which has been suggested to refer to creation of the universe, God said:

> *'Let there be light,' and there was light.*
> *God saw that the light was good, and God*
> *separated between the light and the darkness*
> *God called the light: 'Day,' and the*
> *darkness he called 'Night.' And there was*
> *evening and there was morning.*

However, it wasn't until the *Fourth Day* that God said:

> *'Let there be luminaries in the firmament*
> *of the heaven to separate between the day*
> *and the night; and they shall serve as signs,*
> *and for festivals, and for days and years;*
> *and they shall serve as luminaries in the*
> *firmament of the heavens to shine upon the*
> *earth. And God made the two great luminaries,*
> *the greater luminary* [sun] *to dominate the*
> *day and the lesser luminary* [moon] *to dominate*
> *the night; and the stars.*

It seems to us that the key to Phil's question, *"And where did God come from?"* lies in the interpretation of a unique Supernal Light, other than the light of the sun or moon or stars, being created on the First Day, when once existed only God's creation of darkness. However, we propose a modification to the above statement: *In the darkness prior to the big bang, there were always energy particles, which are God Whose Divine Energy is composed of a unique Mass and a unique Light.*

The great Jewish sage Rashi has said that this Divine Spiritual Light is reserved for Divinely-selected righteous individuals in the future. Just prior to the big bang creating our universe, this Light was confined or was locked up inside these resting energy particles. During the moments before the big bang, God caused His energy particles to release into the darkness or nothingness, causing the big bang explosion. Simultaneously, God's energy particles multiplied as the universe expanded according to currently held scientific theories.

In other words, to answer Phil's question, *God is Light Energy* and has always existed from the beginning of time. As Marcia always said to me, *"God is."* God is the *ein sof or ain sof,* the limitless Supreme Being, and therefore no one or no thing existed before Him. Even the primordial darkness that existed alongside or within God's Light Energy prior to the big bang is God's creation.

In later chapters we shall describe these energy particles of God in quantitative and qualitative terms. God's energy is immense, and tiny "pieces" of His energy particles in the form of "irregular pear-shaped dumbbells" become the energy particles of our Divine souls. We are all thus miniscule "fragments" of God, making us creations in God's image.

If God is composed of energy particles, then it would be natural to assume that His energy particles would behave according to Einstein's famous equation:

$$E = mc^2$$

where E equals the total internal energy of a body at rest, m equals mass referred to as relativistic mass, c equals the velocity of light in a vacuum, and c^2 is a conversion factor needed to transform units of mass to units of energy. One caveat, however, would have to be unique in God's energy particles. The particles, perhaps due to the uniqueness of their M, or Mass, have intelligence. On God's level, these energy particles represent an organizing structure that is the being of us all and our universe. On the human level, the enormously smaller number of energy particles, which are a gift of life from God, are used by God to form our Divine souls. Each soul in turn is a mini brain with intelligence and other human attributes like language and love that we shall discuss in the ensuing pages.

Once the sun was created, it became the luminary that replaced the special Supernal Light of the First Day to light up the day on the symbolic Fourth day and thereafter. God's special Light presently lies in waiting. According to the Hebrew Scriptures, we shall see this special Light again on God's final Day of Judgment in the End Times which shall signal the Messianic Age.

In Jewish thought and mysticism, the Divine soul does not take up space inside the body. If we carry this concept a step further, God would not take up His enormous energy particle space in the immense space of our universe. God is just invisible to us and will make Himself known at the time of the Messianic Age at the End of Days. The Bible talks about us knowing God as the waters of the sea in this new Garden of Eden era to come. As discussed by Dr. Michael Newton in his books, *Journey of Souls* and *Destiny of Souls*, the soul's energy can enter into and appear in transparent human form without the physicality of a body. The soul can also enter animals and create unique butterflies and squirrels. According to our interpretation of the biblical prophecies, God too may appear in human form at the time of His choosing and speak to the entire world.

In Einstein's equation, the 186 thousand miles per second velocity of light, c, is approached by an object when its energy and relativistic mass are essentially infinite. Because God is limitless and beyond infinity and because we are a part of God, His energy and the energy and our souls, one and the same energy, is transmitted telepathically, as we shall discover in later chapters, at speeds of 186 billion miles per second. Incredible as it sounds, energy follows thought and thoughts as energy are transmitted one million times faster than the speed of light. Marcia's soul is 93 billion miles away in Heaven and her thoughts travelling at 186 billion miles per second would therefore reach my soul in my physical body within one-half a second. God is independent of both time and space so that He would be in Einstein's fourth dimension of time-space as predicted by Einstein's Theory of Relativity.

For God, fourteen billion years of evolution are the same as the 5,773 years since the creation of Adam and Eve. Both time events are instantaneous time for God, although Creation and evolution are separated by billions of years. For our human brains, however, we see these events as distinct time entities. But our souls like Marcia's inside the spirit world and mine inside my brain, made up of God's Light energy, know differently. Marcia's soul in Heaven instantaneously talks to my soul in my physical body, because the silent telepathy of our thoughts hitches a ride on God's invisible special electromagnetic Light waves of our energy particles travelling at incredible speeds one million times faster than the velocity of light. This could never happen with ordinary conversation because sound travels at significantly miniscule slower speeds than even regular light. We shall elaborate on this energy link of the spiritual world to the physical world later on in our story.

Marcia and I see our Creator as the Master Scientist Who created the universe and all beings, so that scientists could seek answers in their scientific hypotheses and experiments to understand the functioning of the human body, the cosmos, and nature. We believe that God created a dynamic, interesting world full of wonders intended to stimulate the inquisitive, curious, imaginative mind like that of Albert Einstein. God gave us free will so that especially disinterested parties or atheists would have the choice to doubt Him and even become anti-God.

God gave us intelligence, creativity, wisdom, language, and laughter, along with free will to morally decide for good versus evil. In the following pages, we would like to explore the contributions of our souls versus our brains and bodies. What role does the soul play in intelligence and emotions, compared to the human brain? When do they interact in concert and when do they act independently? How do bad and evil come about when we are faced with moral choices? Is the Devil alive to influence us? Do evil people have a Divine soul? Is the soul responsible for thought, and if so how are thoughts transmitted or picked up? Does thought follow energy or does energy follow thought?

How does the psychic or necromantic make contact with the spirit world? Back when Marcia and I were living in New York, I was looking one evening for something to watch on television. I flipped the TV channels and randomly stopped at *Larry King Live*. His guest that night was an attractive, middle-aged, British woman, who claimed to be a necromantic, a person who has the power to contact the dead. She was taking phone calls, and reaching trancelike states to connect with callers' dead relatives. One of the phone calls was from a mother who was grieving over the loss of her infant son. After a few moments in a sublime meditative state, the British woman made contact with the mother's son, and began to open and close her hand in the direction of the TV screen, repeating the word...*love...love...love.* At one point in the show, I chuckled as Larry King frantically waved his hands in the air above, and asked, "Where are they?"

He was trying to touch the spirits of the dead—of course, to no avail. I reflected on what I had observed, and spoke aloud, "That's neat." I went to bed, and thought no more of this mystical TV episode—that is, until the surprise of the next day.

At the appointment at my Long Island holistic doctor's office, I found myself talking with a woman, whom I had previously shared conversations with while we were each hooked up on our intravenous lines. I was into this nutritional stuff beginning in the early 1990s, and I had all of my silver mercury amalgam fillings in my teeth replaced with porcelain. As a scientist I knew that a lot of the mercury would be ingested during the dental procedures, and thus I elected to have intravenous treatment to remove mercury out of my body. It took twenty-six weekly treatments, followed by oral chelation, to eliminate all the mercury in my cells, as carefully measured by blood and urine analyses.

The woman next to me, to whom I was speaking, was having EDTA chelation treatments for her heart. She paused, and then asked, "Did you happen to see *Larry King Live* last night?"

"As a matter of fact, I did," I replied.

The woman then asked, "Do you remember the clasping and unclasping of the hands, and the repetition of the word *love?*"

"Yes," I responded enthusiastically.

Then the surprise. "That little boy was my grandson. That's how he motioned all the time, and the caller was my daughter. We have been sad for the longest time because we lost such a special child filled with love."

I was speechless. This woman was a nice person and she had no reason to lie. I realized this was no coincidence. I was meant to be a part of this unusual happening. I was being shown something quite paranormal and quite remarkable. At the time I thought God had created an interesting, dynamic world in which people were given gifts of contact with the dead, extrasensory perception, the unique ability to connect with ghosts and spirits, past lives, UFO experiences, and so forth. Today I also feel that the same is true in reverse. The spirit world can contact us without our seeking it out. Our loved ones as souls can come back in our dreams or poke us or move pictures around or show us visions of themselves or create animals like the unique butterfly I observed at Marcia's grave. It is all done through the principles of energy as suggested by Einstein in theoretical equations and proven practically in scientific experiments.

How is God involved? How many energy particles existed in God just before the big bang, and how many is God made up of now? How many energy particles does it take to make a soul, and do we all start out the same? Can we acquire more or less energy depending on our moral choices? Is God the scorekeeper of our good and bad choices, or is God, like Einstein depicts him, a God who doesn't interfere or interact in our lives or doesn't judge us for every moral act?

Einstein's God is only responsible for our universe(s) and its harmony. Einstein believed that all is predetermined and there is no free will. He therefore did not believe that God was our puppeteer. But then how can we explain past lives and trips to the spirit world like the ones that I experienced under hypnosis when I visited Marcia? How do we obtain answers to all of these and so many other questions? Are Marcia and Jerry, your authors, just blowing smoke and all of this to be spoken about absurd and preposterous? Or can we really use the energy of our Divine souls to answer many of the questions worthy of answering?

2

BUTTERFLY

I RECENTLY COMPLETED MARCIA'S MEMOIR, *THE WING OF THE BUTTERFLY*. The title of the memoir came to me from a love poem I wrote and read to Marcia at her gravesite, ten days after she died.

Marcia, you are the colors of the rainbow all in one
You're the morning breeze and the setting sun
The snowflake on the whispering tree
The deer and rabbit scampering free
Your heart is the gentle beat of the butterfly wing
Your voice, the symphonic sound of birds that sing
I smell your breath and see your smile
And know I will see you in a while
I crave your touch and feel your heart
It's not the end, it's just the start
Go take your rest and fly away
But do come back another day
Our souls are bound, full of glee
True love for all eternity

Marcia is the Wing of the Butterfly, as her soul can often be seen as the yellow and black monarch butterfly spreading her wings and hovering for extended periods of time close by her gravesite at the Eternal Light Cemetery in Boynton Beach, Florida. However, at least some of the times I saw a monarch butterfly, it would have to be considered unique; for when our daughter Melanie and I were together at the gravesite, Melanie noticed that where the wings met or divided at the posterior were two large, blue circles, one on each wing, and beside the blue circles were two smaller red circles. The blue color was a beautiful cyan blue, and the red color was like the red ball that the sun wondrously takes on as it sets to end the light of the day.

Marcia's soul not only became the butterfly. Our cousin Roberta sees Marcia's soul in a unique light-brown or gray squirrel with a burnt orange red crest of hair sticking up and running from the back of the top of the head down the back of the body. Roberta has visualized the squirrel at her home and has seen another squirrel looking similar in appearance miles away at her place of work. I have a name for Marcia's soul. She is called *Shubella*, which means beautiful queen, for Marcia will always be my Queen.

How did we meet? Was it destiny?

3

LOVE STORY

FROM THE FIRST MOMENT I MET MARCIA ON FRIDAY, MAY 14, 1981, WHEN WE WERE BOTH FORTY YEARS OLD, I EXPERIENCED BLISS. She was officially separated from her husband, waiting for her divorce to come through. I wasn't far behind her in my separation. In the singles world, people are much more willing to communicate with you and everyone has a story to tell. A group of single males knew of my separation and befriended me and took me under their wing. It's no coincidence that Marcia knew one of the guys. His name was Burt, and he told me that he was having a house party and that he was short of guys. How could I refuse, particularly since I had purchased my first hairpiece to cover up my balding head. I finally felt that I was ready to be a part of the dating world. Burt told Marcia that he was short of girls, so she too fell for his clever line. It's a good thing I had the piece because I found out subsequently that Marcia did not like men without hair.

Marcia had a lawyer suing her husband for divorce. He soon became my lawyer when it became clear that mediation between my ex-wife and myself was a failure. Marcia would have stayed in her marriage, like I would have, except both my ex and her ex wanted out at the time. I shall spare you the details of what transpired in our first marriages, because we found the rainbow and the pot of gold when we found each other. After I met Marcia, I knew that I wouldn't have to settle. I had found my princess, my queen, my beautiful goddess.

I was introduced to Marcia just inside the door of Burt's house as I was standing talking with a male friend. Before I could say hello, she wandered off and then so did I. I thought her actions were strange at the time. She left abruptly when she heard my name. In a short while, I found myself talking to a tall blonde chick, and after a time Marcia came in and sat on an adjacent couch. Maybe that was the cue for the blonde to get up, because she immediately excused herself, claiming she had to go to the bathroom. I'll always believe that the blonde found me dull and went to check out the rest of the action at the party.

My first words to Marcia were, "You look tired."

She jumped down my throat. "Of course, I'm tired. So would you be if you worked all day, took care of two kids as a single mom, and went to school at night taking courses toward a masters degree."

Wow! I walked into that. "My name is Jerry."

"Mine is Marcia," she responded, but somehow I heard Marci and she didn't correct me the whole night long. I only found out by chance the next night on our first official date.

It was a remarkable party evening, and we talked the whole night long about the singles world, our lives, our values, our kids, marriage, love, and what we wanted out of life. At one point the blonde came back, but she was smart enough to realize that Marcia and I had clicked and she took off. Besides, we simply ignored her. I marvelled at Marcia's philosophy about meeting the right guy. She was still that girl going on sixteen determined to find true love. She knew what true love was supposed to feel like, but I had no clue until I met my Marcia Kay.

14

She said, "If you want to meet Mr. Right, you have to be like the salesman trying to sell his product in hard times. When times are tough, the salesman has to see more clients in order to maintain his salary." She figured that she would have to meet a hundred guys in order to meet five potential soul mates; and out of the five, one had to be her Prince Charming. I hardly got my feet wet in the singles world, but Marcia had at least gotten her body wet. She had been to parties all over Long Island, but her sole purpose was to meet Mr. Right. In fact, I would months later tell her that she walked with a pickle up her rear because none of the guys she met ever got to first base with her except one. Thank God he turned out to be a married cheater.

The country had just come out of the Carter years and the fears of the energy crisis where we stood in long lines for gas on alternate days. Burt had bought a coal stove, which became popular during this period. The night was winding down, and Burt came over and dragged me off to see his purchase. I told Marcia, "I'll be right back." While I was hearing Burt's spiel, I caught Marcia leaving the party out of the corner of my eye. I ran over to her and uncharacteristically for me said, "Marci, you weren't going to leave without giving me your number?" She showed me a teasing coy smile and nodded her head that indeed that is what she was exactly going to do. As is my custom, I carried a pen in my front pocket, and I wrote down her phone number on a scrap piece of paper. I think I might have clutched her hand or given her a peck on the cheek. I don't remember. She smiled and left.

Next day, Saturday morning, I sat in my Long Island Poquott rented cottage trying to decide whether to call. Finally I got the courage to pick up the phone and dial. Thank God, responsible Erin at age seven answered in her tiny voice. If it was Kenny at twelve going on thirteen, she probably wouldn't have gotten my message because Kenny was absentminded and to some degree still is, often misplacing his wallet, his car keys, or his cell phone. Marcia was with her next-door neighbor and they were discussing last night's party. Marcia later told me that she told her neighbor that she had met a man of quality.

Marcia called me back soon after she got my message and the rest is history. We made a date that May 15, Saturday night. I picked her up at her Ten Westminster Court home in Nesconset, and she was wearing this strapless light-red knit top and white shorts. The sweater was more like a bikini top, and I immediately felt stirrings that I hadn't felt in a long time. Marcia could pass for Spanish because of her natural olive tanned skin and she looked spectacular. I was wearing long pants when I picked her up in my van. The night was unusually hot for the middle of May, so we decided to go to my cottage first so I could change into shorts. It couldn't have been a better move, as I found out later that her neighbor had warned her to make sure I wasn't married.

As soon as we entered the cottage, I introduced her to her first drink of Bailey's Irish Cream and put on the music of the Carpenters on my CD box while I changed into my shorts. I remained a gentleman throughout the night as we drank pina coladas and strawberry daiquiris on the terrace of the now-defunct Ferry House restaurant overlooking Long Island Sound in Port Jefferson. I was still calling her Marci, and it wasn't until she showed me some family pictures late into the night that I noticed her name on one of her identity cards. She smiled when I called her Marcia. We drank and laughed and talked the night away, and Marcia was so spectacularly beautiful. She was wearing one of her special perfume fragrances, Shalimar, and my mind was dizzy from the alcohol and her aroma. I loved listening to her voice and the night just seemed to fly by.

That second night together, we became soul mates for the eternity. I told Marcia of my intention to enter primal therapy and she was totally supportive. Marcia was so bright that after she went back to Stony Brook University to complete her bachelor's degree, she applied and was accepted into the Stony Brook Doctoral Clinical Psychology Program, which is to this day still an extremely difficult graduate program to be admitted into. She however realized that she needed to be a mother to her children and concluded in her usual logical way that being a teacher would offer maximum time as a single parent. Marcia had therefore enrolled in a master's degree program in special education.

We were the last to leave the restaurant, and I drove her home. As I walked her up to the door, I took her in my arms and kissed her on the lips, all the while being steeped in her fragrance. I learned later that if I didn't kiss her, she would not have been a happy camper because no matter how great our conversations were, we needed to ignite our relationship with passion and romance. I'm glad I was so uncharacteristically brave, because kissing Marcia was more delicious than the orgasm that resulted from the kissing and followed when we made love. And if you combine kissing her with having an orgasm with her, no words can describe the soaring erotic feelings within me. Her kisses to this day are a rapturous aphrodisiac even now as I dream about having her in my arms.

One night while still single, I was out with my friend Howie, and he asked me, "What is it about Marcia that attracts you physically to her?"

I had never entertained this question, yet I answered him immediately. "The five senses."

Howie then responded, "I don't understand."

"It's simple. I'm stimulated by her five senses. From the very first, I love hearing her voice. I love gazing at her beauty and imagining her in my mind. I love touching her skin and her touching mine. I love tasting her kisses, and I love smelling her."

I continued, "Don't get the wrong idea about the smell thing. I'm not referring to her smell when she wears delicious perfume or her body odor, which she doesn't have. It's when I kiss her and the most incredible smell comes out of the breath from her nostrils. The smell of her breath is an aphrodisiac and it sends me into orgasm."

"Wow!" Howie responded. "I think it would take a thousand people to get the answer you just gave me."

Marcia and I spent the next thirty years together, and except for the years I suffered with bipolar disorder in the nineties, our time together was incredibly special. When I found her back in 1981, I found life. She was all that I ever needed to make me happy. Her memoir, *The Wing of the Butterfly*, describes her remarkable story while she was alive for her sixty-eight-plus years in our earthly world.

I expect to see Marcia in this physical world again when her soul joins with her buried body in God's promised Resurrection. According to energy measurements of our Divine souls, which allow us to measure health in the present or in the future Messianic Age, those individuals who have led a moral, virtuous life shall return from the grave in a healed state. Marcia will be with me once again and will be more beautiful than ever and free of the liver cancer.

4

LIVER CANCER

S BEAUTIFUL AS THE PUFFY WHITE CLOUDS ARE IN THE SOUTH
FLORIDA SKY, THAT'S HOW UGLY CANCER IS. Marcia was
never one to go to doctors, and when she finally did in
early February of 2011, the tumor in the right lobe of her liver
was massive, containing almost a half a billion cancer cells. There
were also three smaller tumors in the left lobe, and the cancer had
metastasized to the bone in her left hip. Within six weeks she was
gone as her liver went into failure.

Marcia was and is a righteous person, and I believe now that
her death was in the hands of God. The cancer just made death
inevitable. She also knew that God had blessed her with extra
years of a life of happiness and joy. Marcia was telepathic. She
saw into the near future. But who could place that talent in her
except for God? God came to her many years ago and foretold
the future for her. She knew not only that her death was coming,
but also when it was coming as the time approached. She wasn't
herself that last year and a half. She had significantly less energy,
and her mental acuity was blurred from the ammonia caused by
the liver cancer.

So what does a very, very smart person who is very, very seldom
wrong do? She takes her soul and departs before her decaying

body is extinguished. I feel certain that the great majority of her soul left her body on that last night, right after I saw that tear coming down her right cheek as I was telling her to let go. The body was then on its own to die in peace. That last tear was all Marcia needed to give up fighting in her usual way. It was almost as if she was waiting for me to grant permission to her to leave during that last moment of intimacy we shared together. It was never in Marcia's nature to give up.

I keep trying to think of words to encapsulate Marcia's magnificent human qualities, but she is indescribable. Marcia was the genuine class act, impossible to duplicate because there will only be one Marcia Kay Pollock. I found true love and life itself when I met Marcia. I'm indeed a lucky fellow for the timeless years we spent together. She is my gift from God, my *bashert*— my destiny.

Marcia died on March 18, 2011, and I wanted to die with her. I had lost my soul mate and the sadness was overwhelming. We were in our twenty-eighth year of marriage and were together almost thirty years. Yet even after her passing, I knew that I needed a way to contact her. By early June I was visiting a hypnotist and experiencing a past life followed by meeting Marcia in the spirit world.

The following is my account.

5

⁕

SPIRIT WORLD

With Marcia's passing, I was desperate. I needed to do something to ease the pain of her death. I never expected Marcia to die, and we never spoke to one another about what we would do if either one of us predeceased the other. We did pledge our love for all eternity and we told each other that we would not remarry because forever means forever.

At the time I did not know that anyone had traveled to the soul world, so I started researching past lives. Years ago I had read a book about a woman who described her past lives with uncanny accuracy. Then on the suggestion of a friend, I picked up two books by psychiatrist Brian Weiss, *Many Lives, Many Masters* and *Only Love is Real*. In his books Weiss talks about souls being allowed to come back if some agreement is left unfulfilled. I believed when I read this statement that Marcia would have to return to me because she had made a commitment to be with me forever. I would speak to her aloud and in my thoughts, telling her that she needed to honor her covenant with me, even though she didn't respond at the time.

I called to make an appointment to see Dr. Weiss, hoping I could become his patient, but I discovered that he was no longer in private practice. His office staff recommended contacting

Raymond Moody, who had co-written a book, *Reunions*, which described a methodology for connecting visually with your loved ones in the spirit world. Dr. Moody's strategy is based upon ancient teachings of making contact with spirits through mirror gazing. I did call Dr. Moody's office and spoke to his assistant. However, I reconsidered visiting him after thinking that this wasn't the technique I was searching for. I didn't want to meet just any spirit belonging to my deceased family members, although that would be something I would consider down the road. I wanted only now to meet Marcia.

It wasn't until May 22, 2011, when I was in a Barnes & Noble store in Boca Raton, Florida, that my heart started racing with excitement. The books I was perusing were written by psychologist Michael Newton, who had treated over eight thousand clients. The clients had gone through past life regressions under hypnosis and had died in those past lives. Subsequently their souls rose as they observed their dead bodies below and made their way to the spirit world.

After feverishly reading *Journey of Souls* and beginning *Destiny of Souls*, I felt I was ready to contact Dr. Newton. I went to his Web site and noticed that he had certified individuals in different states. One of these names came up in South Florida and I called Jules. I remember the date of my appointment because after Marcia died, I kept a diary for several months so I could tell her about what was going on in my life. It was June 2.

Jules instructed me to watch a relaxation video on his Web site in preparation for my visit. He told me to bring a pillow and a sheet to cover myself in case I got cold. While I was under hypnosis, Jules was writing a portion of what I was reporting to him in longhand. He said he would touch my hand when it was important that I should remember a specific moment or incident during or after the session, and he would also gently tap on my forehead to either jog my memory or get me back on track if

I began analyzing and commenting on what I was visualizing. What was very important, he stressed, was not to think about what I was seeing or hearing. I was to only report on what was happening without making any judgments.

Jules mentioned that in about 20 percent of the cases, people were not able to have a successful experience. That immediately worried me as I wondered if I would be one of the failures. I had never been put under hypnosis, nor did I think I would be a candidate. During my honeymoon back in 1965 with my first wife, we had gone to the Pocono Mountains in Pennsylvania. One day the entertainment was a hypnotist. He was able to hypnotize a room full of people, but both my ex-wife and I did not go under. As I approached Jules's condo, I kept thinking about meeting Marcia. Jules had asked me to prepare a list of questions that I could use once I was in the spirit world. However, as it turned out, most of what came out of my mouth was spontaneous and did not follow any sort of script. There was one exception. The one question I wanted to address with someone capable of answering me was when was God going to resurrect Marcia and return her to me in the physical world?

When I arrived, Jules immediately put me at ease and took some twenty minutes to explain the ground rules while we made chitchat. I gave him $500 cash upfront. Then after about another twenty-five minutes of body and mind relaxation exercises that went from the head down to the bottom of the feet with my eyes closed at all times, we began descending a stairway of sixty-nine steps, one for each year of my age…

June 2, 2011

On the twelfth step, I am twelve years old. I describe my house where I grew up in Toronto. It is grey with a black front door and a brown banister separating us from the house next door. The banister leads two cement steps upward to a narrow porch, which is blocked off from the neighbor's porch by a solid piece of wood over six feet tall by four feet wide. Upon entering the house, I immediately find myself at a second door, which I open. Before me appears a long hall, which narrows because to the right a brown wooden staircase leads to the second floor. To the left of the hallway are doorways, the first leading to my father and mother's bedroom and the second to a smaller room shared by my brother and me. Norm is two and a half years older than me, but when I enter through our door, no one is there. On the inside a mauve curtain separates our room from our parents' room.

Our parents' room has a 1950s black and white Motorola television perched near a window looking onto the street. I don't remember the size of the TV but it was pretty big in my eyes. On Sunday nights my brother and I would watch from our bedroom. Jules asks, "What was your favorite program?"

I replied, "An opera program came on at 7:30 p.m. The music touched my soul. I remember the theme, 'If I could tell you how much I love you…' I remember watching Ed Sullivan, Bob Hope, Red Skeleton, Jimmy Durante, and Jackie Gleason. Friday night was reserved for boxing sponsored by Gillette with its nifty advertising tune. Every Friday night, we were joined by my mother's black Avon salesman. I still remember his face."

Against the wall adjacent to the hallway is a three-drawer brown dresser, which my brother and I share. Jules asks, "What was your favorite piece of clothing?

I respond with no hesitation, "None." I continue to describe the room with its uncovered hanging light bulb that bothers my eyes, especially when I have one of my frequent headaches.

Jules then asks, "What was your favorite thing?"

Again I respond, "There was no thing." Then I utter, "My dogs, my cocker spaniel Duchess and my black and white fox terrier Pepper."

Jules then returns to the staircase and we descend another five steps when I am seven years old. I am small for my age, even at twelve, as I didn't shoot up until I was thirteen after my bar mitzvah. In front of me at the end of the narrow hallway is the kitchen with its wood burning stove with a black stovepipe leading up toward the ceiling. The round kitchen table has yellow and black covered chrome chairs and to the left of the table is the sink with a red drip rack. On the opposite wall is an ice box sitting on a dingy, rusty, colored linoleum floor.

I describe my favorite meal night on Sunday nights, when my mother made spaghetti and meatballs. The meatballs were small, and my mother simmered them all day in the sauce. I remember my father running a penny ante card game with four of his co-workers at the kitchen table with the men drinking beer from the icebox. My father was the dealer and took a percentage of each pot while I snuck out of bed to watch them play their different card games while bantering back and forth.

The second floor was rented out to a Ukrainian family with two daughters, one brunette and one blonde, who I thought even at a young age were beautiful. As I got older, I fell in love with the brunette, Olga, as I masturbated furiously thinking about her. The family had a kitchen and a bedroom, while the middle room was reserved for a kind Italian gentleman who wore heavy gold chains around his neck and wrist. The third floor was really an attic where one of my father's co-workers lived. He was drunk most of the time, and many nights I would hear him staggering up the stairs.

I descend further in time to the fourth step when I am four years old and am in constant pain from headaches. Then even further as an infant remembering my mother playing with my penis. Then wham! I seem to be floating like I am dead and I am seeing pictures of a foetus. I am rocking back and forth and then I am doing the dead man's float. I speak loudly. "I hear it."

Jules asks, "Hear what?"

"I hear my mother's heartbeat." [I am in my mother's womb].

The floating stops and I'm being rapidly propelled down a dark tube. Every little movement seems to frighten me and the movements are quick as I bounce from side to side as if I'm on a vibrating roller coaster. Then terror!

I feel my head and shoulders being crushed as I slither toward a light. I can't seem to move. I'm stuck and panic sets in. Jules asks me to go back to when I was floating. I do and see another figure floating close by, my twin? We mimic each other in Tai Chi movements. Then I see the other figure disappearing, being pulled by a powerful force that is also taking me with it. I reach out. Something stops me and I'm pushed backward in the opposite direction. Then silence! It's over. My back feels ice cold. I can't warm up.

The scene switches and I see a light in front of me. And then clouds of grey floating in grey-blue space. There's a baby in the clouds. It has blondish hair and now it's disappearing. I see an eye looking at me, then two eyes. Wow! I see my face in the clouds and then my brother Norman as an adult. I see Marcia's face trying to break through, then members of my family.

I am back somewhere [birth room in hospital] *feeling very tiny with my eyes closed. I feel all wrinkled up and ugly. My body feels disjointed and twisted. A light from above is blinding my eyes. Someone picks me up and wraps me* [in a blanket]. *I feel warm. I'm carried over to someone* [my mother] *and I hear the words, "Get that ugly baby away from me." Someone is holding me* [nurse] *and is comforting me. I feel like I'm flying, yes, I'm flying with this person holding me. I see someone* [a man] *looking down on me and I'm spinning A man in a funny outfit reaches down to me and then all disappears.*

[Past life] *It's dark, raining, and cold. I'm holding my hands around myself shivering. I'm on a ship and a cliff can be seen in the distance. The sky is dark and lightning flashes in the air. The ship is rocking back and forth, and I'm barely hanging onto the ship's railing in the storm. Two other men are on board, but I'm a boy perhaps thirteen or fourteen years old. I have a handsome face and my hair has blond curls. I hear one of the men* [captain] *calling me "laddie."*

The storm passes and all is calm. Jules asks me, "What year is it?"

I say, "I don't know." I feel a tap on my forehead. I respond, "One." Then another tap, and I say "Seven." The last two taps bring on a "Four" and a "Six." It's the year 1746.

Jules asks me the name of the ship.

I reply instantly, "Elmira." I feel that I'm English or Irish, and the man I see in front of me seems English. He has a moustache and is smoking a pipe. Later we are eating and both men are drinking whiskey.

Jules asks, "What is your name?"

I reply, "I'm David." "David who?" he asks.

First I say, "Smith," then "Samuels. I'm David Samuels Smith." One man reminds me of the boarder who rented out our attic. The other man [captain] reminds me of Long John Silver. He has curly black hair to match his thick black moustache.

[Soul departs body] The ship crashes into large rocks. We are all thrown overboard into the water. I go under and I try desperately to stay above the surface, I sink down under the water to the bottom, drown, and I then I feel very odd. I am rising [the soul of David Samuels Smith]. I look down into the water and see my dead body. As I rise above or rather am pulled up beyond the surface, I look back again to see my dead body lying very peacefully. As I rise, I see a figure [soul guide] above me encompassed by bright light holding his hand outstretched. Our hands touch. I see his face but it's no one that I recognize. I hear my name but it's not David. "Hello, Jerry, come, come, come with us." There are no steps but I seem to be walking upward. Then I hear a deep voice saying, "Come."

[Meeting Marcia] And then it happened. All that I had hoped for, dreamed about! Marcia's face and head and partial bodiless body all engulfed in brilliant light appear, and she is smiling at me. I look up and she now seems to be crying. I put my arms around her and she is looking directly into my eyes. I don't feel her physical body, but I can "touch" her face and dry her tears. I feel a light touch on my face and we kiss each other non-physically. She is so beautiful, so beautiful. I hear her melodic voice say, "I love you." I'm crying while repeatedly telling her, "I love you." She is bathed in pure white-yellow light and we kiss once more. Marcia drifts off as if someone is taking her away from me. I can't keep her from leaving me. There are no good-byes.

The guide appears. He takes me upward. He has a beard and grey hair. Jules asks, "What is your guide's name?"

I respond, "Elviron."

"And your name?" Jules asks. "Sagittarius" comes immediately out of my mouth.

We land on a small white marble platform. We pass through cloudy areas that appear as white-bluish. I see a picture of my past life. It's David Samuels Smith. Oh, my father and brother float by. Elviron takes me by the hand and we float further upwards.

Thank God! Marcia again reappears and takes my hand, and we enter a bright area with rocks in the water and a beautiful waterfall. I see green trees and forest and then a canyon. We just seem to be drifting in this beautiful place. Elviron is there as well. Clouds appear and everything becomes blurry like in a sandstorm or snowstorm. All is gray and I see Marcia's face. Then I see figures that appear to be tiny people but they don't quite look human. They seem so happy and are dancing in a group [a soul group]. As we approach, they are more like transparent, misshapen, oblong dumbbells.

[Meeting God] Oooh! The light is very bright, exceptionally bright. I keep repeating how brilliant the light is. I am squinting and keep rubbing my eyes. I hear a distinctly deep voice. "I am Who I am." Then he calls, "Jerry."

I respond, "Yes, Master."

He speaks, "You have come. Have you come for Marcia?"

"Yes," I say.

"You cannot have her because she is Mine," He says.

I say, "Marcia belongs to me."

He asserts, "You shall have her when I say."

I boldly state, "I want her now."

He quietly says, "It is not the time."

I plead, "When?"

He utters, "I will tell you." He closes with, "Good-bye, Jerry."

[Meeting Moses] *All the brightness fades and I find myself descending with Marcia, but then we go up again and move sideways through space. I see a bunch of older men at a table and my eyes are fixed on one of them. He* [Moses, I think] *is all lit up and he is carrying a wooden staff that is perhaps six feet in height. There is a wider bronze design on one end. He is wearing a long robe and sandals. I ask, "Can I hold your staff?" He hands it to me, and then I find myself walking up a mountain but Moses is no longer with me. I reach the peak very quickly.*

A booming voice [God] *calls out, "You have come," as if I were expected. I seem to be in my twenties or thirties and am clean-shaven. I feel secure holding the staff. The scene changes as I find myself off the mountain.*

[Taken to my soul group] *A woman in bright light green addresses me, "I shall take you, Moses." I am taken to a group of intensely bright souls. Oooh, they are so bright, brilliant white-yellow, like white gold with barely a separation of the light* [souls]. *Rather, the light coalesces together. It feels like nothing I have ever seen before. I have to close my eyes as it's like looking into the sun, the sun's rays. I feel my heart opening up like the petals of a lotus flower to the light shining down upon me and radiating inside of me. The light enters my heart, and I feel the light in my left arm and left foot. The light force continues to rapidly move throughout my body. I'm all aglow and feeling magnificent.*

I hear that same deep voice say, "Jerry, you are Mine."

Marcia is rubbing my face and kissing me. We have energy between us [light energy]. *She touches me everywhere and it feels like healing energy. I don't feel anything solid, just her light touch. My eyes are filled with tears. I can't stop crying because I am so happy being there with Marcia. I feel her light all around me and inside of me. I hear her sweet voice, "Bye, my love."*

"Good-bye, Marcia." She blows me a kiss and then she's gone.

The session ends. I am back in my physical body waiting for Jules's cue to open my eyes. I'm crying profusely thinking about

Marcia. I love her so and feel so grateful for having seen her in the soul world. It's four hours to the minute since we began the session. I tell Jules that I was hoping I would see Marcia, but I never expected the session to go so well. It was pure bliss being with her.

We continue talking and make plans to do another session very soon. I mention to Jules that I am having temple pain and wonder if he knows of any way that I could treat the pain. He begins to describe energy measurements and tells me that we should do a session to correct for physical and emotional problems. He copies energy materials for me to read and hands me a copy of what he had recorded during today's session. The cost of the next past life-spirit world session would be $300, and he offers to teach me about energy measurements for $200.

Please note that the words written above in the square brackets of the spirit world session are interpretations by myself or Jules or the both of us. After the session I went back to reading Michael Newton's second book, and it was there that I found a description of what other clients had visualized when they saw souls dangling in soul groups. I too early on saw a soul group like this; however, the brilliance blinding majesty of my own soul group was never described by Newton in his selected patients that he used in the writing of his books.

In the next chapter, we shall summarize what Newton incredibly discovered about souls, but next my second journey experience to the spirit world, which I am sure you will find bizarre, totally unbelievable, and not credible. Jules followed the same pattern of relaxation as previously described with the sixty-nine steps, but I shall skip childhood and proceed right to the intriguing events in my mother's womb.

June 30, 2011

I seem to be once again floating as if I'm dead. With my eyes closed, I see pictures of a foetus. I'm floating and rocking and listening to my mother's heartbeat. I see light but I feel abandoned. I sort of see myself climbing out of water, but I can't see my face. Then, wow, I see a man. He is wearing a white fluffy shirt like a tuxedo shirt. I feel ugly and shrivelled up. My face feels ugly and I'm suffering. I hear someone say, "You're supposed to suffer." I wonder why? He says, "It's your mission." I think, What mission? He seems to read my thoughts and says, "You'll see. I'm your soul, Sagittarius."

We begin to play patty-cake [inside the womb] *with our hands as he twirls me round and round. It's so much fun and we're kind of dancing together like in a polka. I feel a kiss and he tells me, "Don't worry, it will all be OK." He puts his hand on my cheek and again says, "Don't worry, it will be OK." He's trying to help me and warns me, "We're going on a ride and it will be difficult and we will move very fast* [birth canal].*" I'm shaking and bouncing and being crushed on all sides. I feel the pain in my neck and shoulders.*

I hear my guide Elviron say, "Hold on, Jerry." I don't know what he's doing here with me and my soul in my mother's womb. My head is being compressed and someone is pulling at my arms, pulling me out as I'm squirming to get free of the womb. Then I come into the light and my guide says, "We made it, Jerry." I'm born.

I am placed on some kind of table. A man [doctor] *in a gown and mask over his chin is looking down at me. A woman* [nurse] *scoops me up and wraps me in a blanket. She is gently holding me. I look up and see someone else with wings* [angel] *flying above us. The nurse picks me up and I feel secure. The room* [birth room] *is bright, and the faces of the figures in the room are a blur.*

I am in the arms of the angel and we are floating upward out of the room into space. The angel is female. We come to a man who seems to be examining me, and I'm handed over to him. I'm then returned to the angel's arms and the man disappears. We are back in the birth room and the nurse has me now. She carries me over to my mother lying in a bed [hospital bed] *and tries to give me to her. My mother raises her hand and rejects me. I hear Sagittarius say to me, "It's OK." The nurse places me onto something soft* [crib] *and Sagittarius says, several times, "Jerry, go to sleep."*

I fall asleep and am dreaming about the angel. She has a white porcelain sweet face. She's pretty but not beautiful. I can see her perfectly right now. She is waving her hand like a magic wand, and I feel a protective light surrounding me. I feel warm under the blankets although my back still feels cold coming from a cold womb. I'm trying to calm down.

[Past life] *I'm inside a house sitting in a parlor reading in the light* [candle]. *I'm examining something, perhaps some documents or maps. I'm getting a picture of George Washington* [I'm George Washington?]. *He's in a military uniform and is smoking a cigar while strolling around by himself. He tips his cigar into an ashtray to get rid of the ashes. He seems to be worried and has his hand on his chin.*

Someone just called, "George." It's a woman's voice.

He responds, "Yes, Martha, yes, dear."

I see the woman. She tells him, "Put your cigar out and come into dinner." He puts down something rolled up in his hand and follows her. The scene shifts.

He's on the battlefield, and the scene flashes between the battlefield and him being older sitting outside his house on his rocking chair, smoking his pipe. I see him hunched over and he seems to be sleeping in his chair. But he is not. He falls off his chair to the ground. The woman [his wife, Martha] *comes over and bends down beside him. "George, don't leave me." How strange, I think, Martha Washington looks so much like my Marcia Pollock. George died, I think, and I see Sagittarius rising higher and higher while continuing to look back down at George with Martha (Marcia) crying over his dead body. Sagittarius looks so sad and he rises higher and higher. George gets blurrier and blurrier. Sagittarius looks like me* [Jerry Pollock] *when I was in my twenties and thirties and in the clouds he meets up* [I meet up] *again with Marcia's soul.*

Sagittarius (my soul) and Marcia (her soul, Shubella) embrace in as real a kiss as possible without physicality, an incredible kiss, and we clasp hands and rise together. Marcia appears to be older than me and I seem to be like a boy in comparison to her. She's smiling. Ooops! There's the guide Elviron. He is all in green like an elf or a dwarf. Elviron has a distinctive face, but he's not anybody I knew in life. All of a sudden, his clothes disappear and he becomes all light like Marcia and I.

[Council of elders] *Marcia and I are flying hand and hand, and below us I see houses, fields, and mountains. We land somewhere, a museum, I think. We find ourselves in a tall, ornate, domed space that seems very expensive. We are welcomed by someone with a long pipe. He says, "Follow me," The three of us, the guide, Marcia and I, seem to be walking after him* [But maybe we are flying slowly at a low level]. *We come into a very bright room and in front of us is a long, horizontal, ornate, silver table. We see what appears to be transparent, glowing, bright figures behind the table. I sit in a lit up chair directly facing them. The chair has no arms, just a back. There is a central figure and figures on both sides sitting at the table, but they are so bright that my eyes are burning. One huge top piece of glass spans the table length, and ornate pedestals hold up each end of the table.*

Elviron tells me, "You've been summoned here."

I ask him, "Why have I been summoned here?" The central figure at the table addresses me. "We have questions for you. Do you have questions for us? Ask them."

I ask, "Who are you? What is your name?"

The central figure responds, "We are the seven shepherd princes. You know about us, Jerry. You wrote about us. I am King David. Methuselah is on my far right, and next to him are his father, Chanoch, and Seth the third son of Adam. To my left are Moses, Jacob, and Abraham.

I look at King David and it's almost as if I'm looking at myself. I ask King David, "Have you been resurrected to be with me?"

He answers, "Yes, to be with you." I turn to Abraham, although I can't make his face out.

"When will Marcia be resurrected in her physical body, healthy and vibrant?" I turn to Marcia and see her smiling.

It is King David who answers, "In the Time of the End, Jerry."

I say, "When, when?" He responds calmly, "Have patience, Jerry. It is not the time."

Moses then chimes in, "Only when God will say. It is not for us to say."

"But I want my Marcia. I want her with me."

Methuselah speaks, "It is not the time, Jerry."

I am frustrated and want them to answer me. "How can I help bring about Marcia's resurrection?"

King David answers, "You must do your earthly duty."

"The Messiah," I say.

Then King David confirms, "Yes, I was not chosen, but you are. You've been chosen. You out of all of us must suffer great suffering. Your suffering will bring on the Messianic Age."

I argue, "Losing Marcia has brought on my greatest suffering."

Something or someone [God?] seems to be floating above us. I hear, "All of this will help you to take away your ego. Marcia has been taken away from you because you need to realize how important your mission is."

Marcia comes over to me and holds my hand. I plead, "We love each other. Can you answer me, God?"

God patiently responds, "You shall have her, but not now, Jerry. Not now."

I whimper, "I want her now."

In a commanding voice, God says, "I shall intervene at the proper time. You must do your job."

I state, "I don't know, Lord, what I have to do."

God responds, "I shall guide you, but don't think or plan. Just do."
I ask, "What about the Third Temple?"
God doesn't answer my question. "I must go," He says.
I try one more question. "Will I still have to suffer, Lord?"
The Lord quietly states, "Perhaps." He's gone.
We are all still in the room, and I address my Council of Elders. I
ask, "Will you help me with my mission?" but no one answers, because
something else is happening. Brilliant light shines above us. It is extremely
bright. I don't see anything, but I hear, "We are your guardian angels,
Jerry. We shall help you." At first I see nothing except the light and then
I see a face that looks like Jesus off to my right. He is on the cross. Someone
else with wings comes into view. He looks like Michelangelo's sculpture of
King David. I hear his voice, "I am your guardian angel."
I ask, "Are you King David?".
He responds, "No, I am the angel Gabriel."
I say, "You don't look like Gabriel" [as if I know what Gabriel
looks like].
Gabriel speaks again. "I can take on many forms and faces. This is
how I appear to you as an Adonis. To the righteous Daniel twenty-five
hundred years ago, I appeared with bronze arms and a flaming face."
The Creator comes back. "Jerry, you must do good, be a force for good,
for evil lurks in the world."
I assure Him, "I shall."
He replies, "Good."
King David then interrupts, "These proceedings must come to an end."
Marcia comes over. "Come, my love." Everyone disappears. Only
Marcia is left. She is sad. "I know I must part too. She's smiling at me
and kisses my cheek. "Good-bye, my love." I feel myself descending down.
I land right beside where I am lying down in Jules's room, and then I
move into my physical reclining body. The session ends. Jules slowly asks
me to open my eyes. I'm crying again.

I looked at Jules and we smiled at each other. He said, "Hey,
dude, you thought you wouldn't be successful. I have it all

down. What a trip. You were right there with God. I think you should see my friend Hector in California. He does this stuff at an advanced level." I got up and went to the bathroom, all the time thinking about what transpired. I felt like Montgomery Clift thinking about Elizabeth Taylor on his way to his death for murdering Shelley Winters in the movie, *A Place in the Sun*. I saw Marcia in front of me, so beautiful. I returned to Jules, who gave me one of his knuckle-to-knuckle bumps. I feel in awe of what just transpired in the spirit world.

The reader needs to understand that when you experience a past life-spirit world regression, you are going back as you are today, as the person you are and not as the soul that rose when that person in your past life actually died, David Samuels Smith or George Washington, in my case. In the present, I am not either one of these people, although I have 1 percent of each of their souls. Therefore my reactions and perceptions to events or speech communication are entirely different than it would be for them. I did see George Washington's face in my past life regression, but not the exact face of his wife, Martha. Maybe that's because I am not familiar with Martha's face and I could not conjure up her face under my hypnosis. Also, Martha is my Marcia's past life and not mine. I do not carry that tiny piece of her soul but Marcia does. Nevertheless, Martha Washington's face is distinctly known from historical sources. Therefore George Washington's wife, Martha, looked different to me than who I saw in the past life regression, my very own wife, Marcia.

In the case of the Council of Elders, nobody in the world knows what Moses or King David looked like. Yet I was able to picture them as a blur when Jules hypnotized me. Nor do we know Jesus' physical appearance because there were no cameras or painters back then to record his face so that we could view him today. In the case of David Samuels Smith, the blond-haired, handsome young boy I saw in my regression, I did create my own

36

image of him or rather the image was created for me, but I can't tell you if that's exactly how he looked back in 1746.

I was totally surprised to see Martha as someone who closely resembled Marcia. You can't fake spontaneous visions. It is what my brain saw. Furthermore, this discrepancy doesn't make it any less valid, as one of Marcia's past lives as I said was as the real Martha Washington, wife of the president. Jerry Pollock is in a different body with a different brain than the young boy, David Samuels Smith or George Washington. We shall explain connections to past lives in a later chapter, but for now we should view Jerry Pollock's living soul as different from the souls of the young boy and President Washington, who now reside like Marcia in the spirit world. Yet keep in mind for later discussions that I am carrying 1 percent each of David Samuels Smith's and George Washington's souls within my soul. There may also be others who, like me, carry 1 percent of George Washington's soul. As we shall discover in the next chapter, a soul can split into different bodies in different geographical locations in different lifetimes. The inclusion of past souls is necessary to recognize and have past life experiences. Marcia and I lived different lives in different bodies over the centuries. This is true not only for us, but for all of us who have lived past lives.

I have spent thirty years in primal therapy doing deep feeling regression with my therapist Tracee and on my own. In primal therapy I returned to the womb to previously hear those awful words from my mother, "Get that ugly baby away from me." They explain a lot about how I've felt all these sad years ugly and inferior and easily embarrassed and especially not wanted. It may be hard for you to understand how I could remember these words, but that's what she said when I connected to the exact energy of that moment when I went back in time during therapy and found myself in my hospital birth room.

The baby feels or senses but cannot express his or her feelings because the baby does not have language. The foetus or baby is also trying to survive, so it suppresses the life-threatening feelings at the time of trauma. However, the baby's or foetus's brain stores and records those feelings and words in a memory bank until such time when the person returns to the womb or birth room as an adult. It is at this time when the adult can handle the traumatic memories and assigns language to such memories. Those ghastly words, "Get that ugly baby away from me," were stored in my brain until recently, when I unlocked the vault that has kept them secretly hidden for almost my entire life. I've heard them in primal therapy in my fifties and now in a past life regression at age sixty-nine.

What is fascinating to me is that I made new discoveries in addition to my primal therapy experiences about my early life, when I went through these past life-spirit world regressions. Again the memories were secretly stored in my brain because I didn't have language or visual recall to remember them. Inside the womb, the foetus, like the infant, senses and yet cannot speak. The foetal and infant brain, although not developed like the adult brain, still have all of the machinery to record and store memories like your computer does.

I am completely astounded by the conversation between me and my soul Sagittarius. It really happened and we did amazingly play patty-cake. These special memories were not revealed until now. What a wondrous journey. To see Marcia again is what now keeps me going in the present. Without these special experiences, I might have sunk into clinical depression, which I recall still too vividly from my horrible severe mentally ill experiences in the nineties. As days pass, my love for Marcia grows even in death. I see us being together once more when God allows her to be resurrected, just like He said he would during my past life regression to the spirit world.

What do we know about the soul and the spirit world? Is there a life between lives?

6

⨳

WHAT WE KNOW
ABOUT THE SOUL

MOST OF WHAT WE CURRENTLY KNOW ABOUT THE SOUL COMES FROM MICHAEL NEWTON'S STUDIES WHICH BY NOW HAS ABOUT TEN THOUSAND CLIENTS. In addition, investigators such as Raymond Moody, Brian Weiss, Ian Stevenson, and others have added to our knowledge base. There is also a wealth of information described by kabbalists hundreds of years ago. In later chapters, Marcia and I shall add brand-new information based upon our collaboration across two worlds whereby we specifically use the term *Divine soul* to represent soul energy. We define Divine as the soul being our gift from God. We shall demonstrate that without this gift, there would be no life. We offer a brief summary of current knowledge:

1. The soul enters the foetus anywhere from three months until birth. There is a complete neural integration of the soul with the brain of the foetus. Each brain is different.

2. Souls can divide their essence and be in more than one or in multiple bodies at the same time in different geographical locations.

3. Souls choose the bodies they wish to enter when they look ahead while in the spirit world for approximately the first eight to twenty years of their future life. This is called incarnation into a new body so that a soul may have lived many past lives or none at all. Of course, life has many pitfalls after you get out of your teens, but the soul in the spirit world never was aware of this when it chose the body and its new life. I never knew that I would become bipolar at age fifty. Neither did my twin sons in their twenties. We don't see ahead to some adult body ravaged with cancer like Marcia's, nor to heart disease or to autoimmune disorders such as multiple sclerosis.

4. When the foetus is born and becomes a baby, the memory of what he saw in these early years in the spirit world concerning his new incarnated life is wiped out like having a frontal lobotomy of your brain.

5. One main purpose of the soul's new life is to improve on any faulty past lives by helping its present body lead a more meaningful spiritual life. Errors of the past need to be corrected, although the soul may not be aware of these missteps. Essentially the new soul is to help the body improve its character by starting from scratch and making apparently "new" discoveries.

6. The soul doesn't have physicality because it is energy. It therefore doesn't have that human tactile touch, but you can still feel something such as when Marcia recently tapped me on my left shoulder when I was sleeping sitting up on the couch. The same happened to my daughter-in-law Karen after she had a dream about Marcia. Marcia touched her hand a moment before she woke up so that

Karen would remember the dream. In my experiences in the spirit world, I could feel Marcia's light touch on my face, and when I first saw her, she took my hand and we rose upwards. The latter wasn't like a handshake but we did clasp hands. Our kisses were sensuous but non-physical.

7. The soul is transparent, and if it appears to you in human form, you usually see the soul from the waist up. If you are familiar with the *Star Trek* television series or movies, the soul almost looks like that transition period when Captain Kirk tells his engineer, "Beam me up, Scottie." Captain Kirk appears transparent just before he becomes solid and takes on his normal physicality. You can therefore see souls if they choose to make themselves visible to you. If they don't, you will walk right through them without the realization that they are there.

8. Souls in the spirit world are happy. They have no tension or worries, just a sense of well-being in their beautiful world. They float around and sing and dance together.

9. Souls display colored light encompassing the spectrum from white to purple-blue, although I saw brilliant, blinding white-yellow colors on my visit to the spirit world. When I was in my hypnotic regression, I was hugging and even kissing Marcia. Our light enveloped and surrounded each other as we swirled round and round.

10. Souls live in soul groups. My soul group in my spirit world journey was so blinding bright that I kept rubbing my eyes and was barely able to look at them. I could not clearly distinguish them as distinct entities. They had the

same brilliant white-yellow color that Marcia had, except Marcia by herself was clearly visible to me.

11. Souls can look exactly like they looked when they were living in our earthly world. This feature allows us to recognize our loved ones as souls.

12. Usually when you enter the spirit world, you meet a guide. You can also meet your soul mate, like I did when I first met Marcia.

13. I am not sure if anyone speaks in the spirit world. They seemed to have communicated telepathically with me via the mind. You will see in later chapters in the book that I am freely doing this with Marcia now. Newton states that this transmission is akin to electrical sound impulses. Marcia and I believe that the telepathy is not carried by sound but by a special Divine Light energy, a form of advanced electromagnetic energy to the light energy that Einstein famously described in his relativity equations.

14. When souls in the spirit world travel to Earth to return to familiar surroundings, they use a small percentage of their total energy to do this. And even when a soul incarnates into a new body, a small percentage of the soul's total energy still remains behind in the spirit world. When a person dies in his or her new life and the soul departs the body, the energy of the soul joins up with this remaining energy in the spirit world.

15. Bodies can damage souls, but souls are still accountable for their time on Earth. A soul can have a difficult incarnation if it winds up in a body that desires evil over good. Upon

their return to the spirit world, all souls go through a renewal and healing regeneration period. The body still controls its own free will, but the soul can try to influence the body's choices.

16. Souls appear as patches or dots of light in the spirit world when they have not transformed themselves into human form. They seem to be all lit up hanging in clumps, and they may appear pear-, oval-, or oblong-shaped. On my journey to the spirit world, the souls in a soul group that I saw early on appeared as irregularly shaped dumbbells just suspended or hanging freely in three-dimensional space. They may have been more bulbous, like the shape of a lit-up pear.

17. A Council of Elders meets with the soul upon return from its life on Earth. In many cases, the soul faces questions about its recent life in a body on Earth. In my case, I was asking questions of my Council of Elders. The Council of Elders may vary in composition and number.

18. Clients of Dr. Newton have claimed that they feel the presence of a supreme power directing their spiritual world, but they are uncomfortable using the word *God* to describe this being. I have no such discomfort, as I was blessed to speak to God in my visits to the spirit world. In addition, as I wrote in my book, *Divinely Inspired: Spiritual Awakening of a Soul*, originally published in 2003, God spoke to me twice in 1982 and once in January of 1999.

19. Souls can direct their energy into inanimate objects such as rocks and plants. They can also enter animate beings such as butterflies, hummingbirds, and squirrels. Marcia

visits her cousin Roberta as a unique light brown and sometimes gray squirrel with an orange red crest running from the back of the neck down the back to the squirrel's tail. And especially for me, Marcia visits her gravesite as a yellow and black monarch butterfly with distinct red and cyan blue circle markings on its wings.

20. Souls differ in how old and advanced they are. The top tier, like Marcia, dates back tens of thousands of years and they have extraordinary skills and insight.

21. The memories of past lives are imprinted in the soul. Past life memories can help the soul choose its new life. Before Marcia was diagnosed with liver cancer, she asked me to give her my memoir, *Divinely Inspired*. She read all the chapters, but when I asked her why she didn't read the epilogue and my spiritual and scientific philosophies, she didn't respond. Almost nine months after her passing, when I am now freely communicating with her soul, I brought up this subject again. Her answer was that she knew she was going to die and she wanted to remember everything about me from when I was a child up to the present time. The last chapter, the epilogue, wasn't important to her. Souls do sometimes read earthly books in the spirit world.

22. Recent male past souls may have chosen to be a male in their new lives, but they can also choose to be a female. The soul can choose to be a good person in most of the incarnations, but at some point a soul may want to be the opposite, a really bad person. You can't know love without knowing hate, nor can you strive to be more perfect if you don't experience imperfection. God created human beings

as imperfect so they would appreciate a more future perfect world.

23. The soul seems to be a catalyst for the human brain's activities such as imagination and creative ideas.

24. Souls may come and go when the body is sleeping, when the person is in deep meditation, or when the person is given an anaesthetic prior to surgery. Souls are always close by, but they may visit places familiar to them during this absence. How is this possible if the body can only live with its soul present?

25. Soul mates, whether both are alive, one is dead and one is alive, or both are dead, are never far from each other. Marcia and I now have a love across two worlds. We speak to each other telepathically on a regular basis.

26. Souls are energy — Light energy. Light has properties of a wave and an electromagnetic particle force. This means that the soul can work in mysterious eerie ways with objects. Marcia as a spirit has moved pictures around to her liking in our home. She even has flipped pictures of our daughter and son-in-law on their side.

27. Souls can enter dreams like my daughter-in-law's dream or my own recent dreams of Marcia. She can also enter my meditative state. Marcia put me together with another woman in one of my dreams. When I checked my energy, I realized she had done this to suggest that I see other women to take a way my loneliness. Now that we freely communicate, I scolded her for this and told her we are bound for the eternity. Notably, she has done

a hundred-and-eighty-degree turn and doesn't want me
to see any woman at all for companionship. That makes
sense because Marcia felt this way when she was alive. She
would tell me, "There's no room for three in my bed."

28. Survivors who don't feel the presence of a loved one
    right after death will in the future. However, some may
    never pick up anything. If you truly love someone or the
    departing soul was a soul mate on Earth, then there is a
    good possibility of feeling that departed soul's presence.

29. Only certain people can see the soul as an apparition in the
    physical world. After my second session with Jules, I saw
    Marcia everywhere for a day or two.

30. Newton says that the soul has intelligence and we concur.
    A Divine intelligence put us in this world to learn and
    grow for the greater good. None of us are here by accident.
    We shall develop this concept further when we discuss
    God's plan for humankind.

31. Souls carry all the five senses of humans when they leave the
    physical world, but experience a loss of tactile receptors.
    Marcia and I need to explore this statement further. The
    human brain, for instance, contributes to smell and taste
    and to speech and hearing. We need to separate the role
    individually of the soul and the body and any synergistic
    activity that the soul and body come together on to promote
    the well-being, joy, and satisfaction of the individual.

32. Souls are created but clients with Newton could not explain
    the process. A supreme being was implied from Newton's
    studies. Marcia and I shall attempt an explanation.

33. We find it interesting that when souls choose a new body, they pass through a set of doors into a hallway and then through another set of doors. The hall is the transition point between the spiritual world and the physical world. We would like to point out that there is another Hall, which was part of God's First and Second Temples in Jerusalem and will be part of the future Third Temple on earth. This Hall leads into the Holy Sanctuary in the First Temple and at the back of the Sanctuary is a royal purple Curtain behind which is the Foundation Stone on which the Golden Ark of the Covenant containing God's Ten Commandments sits. The Ark of the Covenant was not present in the Second Temple but it will be present again in the future Third Temple. The Hall in the Holy of Holies is like the hall where souls pass through to incarnate into new bodies. Both hall and Hall are the separation points between the physical and spiritual worlds. You can meet God in both worlds, the physical and spiritual, if you are fortunate and blessed as Marcia and I were and are. In the future Messianic Age, all those who are a part of it will have an opportunity to know God.

34. The soul has the right to leave a decaying body with no possibility of recovery and no quality of life remaining. In Marcia's case, the great majority of her soul left her body seven hours before her body was pronounced dead.

35. A soul receives the mother's emotional feelings inside the brain of the foetus. That's how a soul knows if the mother wants the baby or not. In my first session with Jules, my soul Sagittarius warned me of the tortuous trip ahead in the birth canal of my mother's womb. I was an unwanted baby.

36. The language in Heaven is completely different from our diversity of languages here on Earth.

37. Ultimately the energy of the soul radiates throughout the whole body. This will come up in our discussions when we talk about energy channels. We will also define the initial location of the soul in the brain prior to its spreading throughout the body.

38. Hypnosis can help overcome the amnesia state brought on by birth where the soul doesn't remember any previous glimpse into its future body, Other techniques such as deep meditation, yoga, Tai Chi, prayer, extreme physical exertion, and the imagination can also trigger certain memories that the soul already had. The amnesia at birth allows the soul to make its own discoveries in a new body.

39. The soul cannot express itself through a chemically addicted mind such as in individuals who are users of mind-altering drugs.

40. Some souls remain on Earth usually because they are unhappy with their recent life. You find them as ghosts creating mischief in haunted houses. You may see them and even hear them. Since I have opened my mind up to past lives and the spirit world, I have had many visits to my bedroom by such entities. Some are harmless but some are evil.

41. You don't have to go through a past life to encounter souls or spirits. Dr. Moody's technique of mirror gazing allows you to do this.

42. Different religions have different ideas about the soul and the levels of the soul. Judaism speaks about five levels of the soul. Mysticism plays a large role when trying to understand the soul.

43. Souls are continually learning as they attend classes in schools. They know all about are earthly world and everything from the past, present, and even the future is recorded in libraries and other centers in the spirit world. Knowledge of souls may vary depending upon how advanced the soul is. Advanced souls serve as teachers for novice and less advanced souls. Marcia is helping teach the children who have passed on at an early age, as she was a special education teacher in this life.

44. From Marcia and my collaboration, we know that when a soul enters a foetus, it consists mostly of a new piece of God's energy. For example if a soul has lived twenty past lives, then that soul entering a new body will contain 20 percent, 1 percent for each past life, of souls that have lived in other bodies. For example, I believe that I carry 1 percent of the souls of David Samuels Smith, George Washington and Moses. Recently through soul energy measurements, I discovered that Christian male and female friends carried 1 percent of the souls of the apostle John and Mary Magdalene while I carry 1 percent of the soul of Jesus. If I lived twenty past lives, then the new piece of energy I received from God, when I entered the foetus in my mother's womb and became Jerry Pollock, represents eighty percent of my total soul.

45. Marcia and I concluded that the soul may have multiple purposes for returning to a new body in a new incarnation.

Some souls may choose to return simply because they miss the passion of the physical world. In part the soul always returns to rectify the failures of the past lives it has lived although God may have a larger spiritual plan for certain souls. We believe that our spiritual destiny is tied up with the coming Messianic Age and the End of Days. We say this because we believe that the soul in our bodies influences and often governs our actions and the choices that we make. I started a non-profit Shechinah Third Temple corporation which has everything to do with the End of Days and during these past thirteen years I have spent an enormous amount of time with the ancient biblical texts. You shall learn more of this connection in later chapters.

We really cannot proceed in our understanding of soul and its energy unless we look at the physics of energy through Einstein's eyes. Energy is energy, or is it?

7

⚜

EINSTEIN

MARCIA AND I NOT ONLY HAVE AN EMOTIONAL, PHYSICAL LOVE, BUT WE ALSO HAVE A VERY SPIRITUAL LOVE WHERE NO ROOM EXISTS FOR UNCERTAINTY IN OUR FAITH IN THE CREATOR. For the last thirty years of his life, Albert Einstein unsuccessfully tried to convert the uncertainty of quantum mechanics into certainty by finding a unified field theory to accommodate both the particle and wave properties of light as well as any energy form, including subatomic particles. His belief was that God never rolled the dice so that chance in the universe could not possibly exist.

Given the premise that he viewed God as the Creator and maintainer of the universe, Einstein's philosophy was that the answer was there before the question. He just felt that he hadn't yet found the answer of unifying the concepts he initiated for a deterministic universe. He certainly was persistent in his beliefs as he tried to solve the unifying riddle by deriving complex mathematical equations to prove his theory right up until his death.

If you were to compare King Solomon, who lived twenty-eight hundred years ago, with Albert Einstein, you would probably rightly conclude Einstein was more intelligent. However,

Solomon was probably wiser for what he had to deal with in his time. Einstein's belief in God stopped short, but Solomon's did not. Solomon looked for answers in both the physical world and spiritual world. Einstein on the other hand relied on his own creative genius, believing God did not interfere with or was a judge of human affairs.

Solomon, despite his faults, wished to be guided by God. We don't know how far Einstein could have gone had he allowed God to guide him. He may have been proven correct, or he may have realized that the uncertainty offered by quantum mechanics was all part of God's predetermined plan to create a dynamic interesting world for humankind.

If you think of beginning and destination points on a geographical map, you'll find numerous routes to get to your end point. The great Rebbe Menachem Mendel Schneerson said that the kabbalah provides points on a map, but we humans in our lifetime have to figure out how to connect these points to get to our destination. Einstein was exceptional at finding the exact route through the maze in his brilliant scientific theories. However, it is God Who provides the map and the points on the map for us to explore whether we choose to believe His role in our world or not. If we have faith, He helps us connect the dots. God has His own plan for humankind and whether we get there tomorrow or at some future predetermined or random date is not relevant. Time has no meaning for God because He was, He is, and He will be. Einstein importantly accelerated the pace at which we shall arrive at our final destination. The final destination is not the end, but it starts a new glorious beginning of a second Garden of Eden in the Messianic Age, where we take the next leap in human evolution.

The sages teach that God leads people in the direction they wish to go. Yet Marcia and I believe that sometimes God in a subtle and sometimes not-so-subtle way leads man in the direction

He wishes them to go. The not-so-subtle way was evident in biblical times for Abraham, Isaac, Jacob, Moses, Kings David and Solomon, and Jesus. In subtle terms, God perhaps led Einstein in a certain direction, particularly since Einstein claimed that free will does not exist and was only popularly created so that we would act responsibly and make good moral choices. Most people would not agree with Einstein on his opinion of free will. Yet we all understand how Einstein, the man, could believe in what he believed in, because if all of us went in one direction, Einstein would be going in the opposite direction.

In political terms Einstein would be called a maverick who bucked the norm and therefore invariably chose the road less travelled. He was willing to chuck long-held concepts in classical physics so that early in his career, he was able to use his creative genius to develop unique original theoretical, later scientifically proven, equations for special relativity and general relativity. His work is generally regarded by physicists as the most significant transformational discoveries and contributions to our current understanding of the universe and to so much more of what we now know about our modern world. His equation $E=mc^2$ is probably the most famous equation of all time, known by scientist and layman alike.

In Walter Isaacson's biography of Einstein, we see the first description of light as an electromagnetic wave akin to Maxwell's original discoveries. By coupling the effects of magnets on electric current and in reverse the effects of electric current on a magnetic field, an electromagnetic wave was generated. However the genius of Einstein went further and hypothesized that light comes not just in waves, but in tiny packets as quanta of light later dubbed photons. Einstein was able to calculate the energy of these quanta, but later became sceptical of the same quantum theory he had originated. Virtually on his deathbed, he uttered, "What are these light quanta?" Quantum theory so far has stood the test of time for our universe, but all bets may be off if other universes exist. Quantum theory was further developed and championed by famous physicists such as Neils Bohr and Werner Heisenberg. The problem for Einstein as we expressed above was that quantum theory was based upon uncertainty.

There is a belief in Orthodox Judaism that God's angels supply electromagnetic energy to the sun, moon, and stars, and then from these luminaries energy comes to the Earth and planets. In turn we humans are continuously interacting with this electromagnetic energy of our universe luminaries.

In Eastern cultures, this same electromagnetic energy was and is believed to run through our bodies, and by practicing Tai Chi, Qigong, and yoga, one can absorb this energy from the Earth and the Heavens. In Chinese medicine all wellness depends upon the continuous unblocked flow of electromagnet energy through our brain and organ systems. A myriad of conditions, including the stressful vicissitudes of life, cause the loss of harmony of energy in our bodies, and the consequence is disease or even death. Some Qigong masters have accumulated and stored enormous amounts of positive energy in their bodies, and very ill people have desperately sought out these healers to infuse this positive energy into their own sick bodies. When you see martial artists break thick wooden blocks with their finger, they are bringing the full force of this energy to the task of breaking the wood. The old Tai Chi masters could give you a powerful kick and send you thirty feet across the room. When they stood in Wu Chi standing pose, their energy in their feet would sink one and a half feet into the ground as if they were rooted like a tree.

If you go back to the Creation of Adam in the Garden of Eden, life begins with God breathing His breath into Adam. The breath of course is the Divine soul, which is God's special electromagnetic energy, with the exception as we shall discuss of the soul energy acting as a mini brain. This special electromagnetic energy that circulates in our organs and from our organs to all of our cells we propose is the energy of our Divine souls. Without this soul energy we would die.

God in dramatic fashion did truly breathe a living soul into Adam at the time of Adam and Eve. All of evolutionary humans and animals before and after Adam have souls that have kept them alive these past billions of years. A car like humans has all its parts but does not come "alive" or start until you turn on the ignition. The soul does the same thing with humans. When the foetus is inside the mother's womb, it is sitting in the back seat so to speak while the mother drives the car. The mother keeps the foetus alive through the umbilical cord attachment. However when the foetus is disconnected from the mother, is born as a baby, and must now survive on its own, it cannot until the soul energy inside of it ignites its life. The soul like oxygen is a vitality factor. Without either one of them we die. God acts behind the scenes. He always allows a strictly scientific explanation for his miracles. You can explain life without the need to invoke the energy of the Divine soul.

Einstein defined this electromagnetic energy by his formulas of special relativity and general relativity. We are just putting God into his equations in a unique way. If a Tai Chi master or Qigong healer has powers that can be demonstrated visibly to us, what would be the power of God, who has vast limitless, unimaginable amounts of His electromagnetic energy at His disposal? God is Light and Light Energy, similar or almost identical to the same kind of electromagnetic energy that Einstein discovered. God's massive amount of energy, which is accompanied by a supreme organizing spiritual intelligence, could easily be envisioned to carry out almost anything in our world, including Divine miracles. Moreover, if our Divine souls are but "pieces" of the Creator's energy particles analogous to quanta, then it's further possible to envision how telepathy can occur between a psychic and a spirit or between Marcia and myself, who was taught by Jules to measure this internal unique electromagnetic energy.

There is something scary about all of this besides the natural scepticism that such hypotheses generate. Besides the light of the Divine soul, there is dark energy that can operate under God's full knowledge and purview. God created Satan the Devil to influence the evil inclination and to be a negative force in our brain. However, the Devil has taken it upon himself to act independently of God and to be God's foil and thwart God's plans for a Messianic Age. The joke is that we have been fooled into believing that the Devil is just something out of Greek mythology. However, the Devil is alive and has been since he was kicked out of Heaven as a fallen angel many moons ago.

Satan is extremely powerful, but his dark energy is not based upon light, as his Divine soul is completely darkened. We cannot therefore apply Einstein's energy equation to the Devil since Einstein's equation uses light. We need another way to explain the Devil's powers. In my experience Satan can telepathically connect to me and speak exactly like Marcia. He has done so with me on many occasions. Each time I must check whether I am speaking to Marcia or to this evil creature. The Devil also can incarnate into human beings, and we believe that he has descendants, as I described in my recently published novel, *Gog and Magog: The Devil's Descendants*. We shall have much more to say about the Devil owning peoples' souls in a later chapter, as wild as it sounds for the reader hearing about the powers of the Devil for the first time.

Dr. Henry M. Morris, who is recently deceased, wrote about the prophecies of the Wars of Gog and Magog as referenced in the New Testament, Revelations 20:7b-8. "Satan shall be unloosed out of his prison to deceive the nations which are in the four quarters of the Earth, Gog and Magog, to gather them together in battle, the number of them as the sands of the Earth." Dr. Morris believed that Satan came out of his prison at the beginning of this new millennium. We would add that Satan is preparing his cadre of human demons, whose souls he owns to counteract God's forces of good.

Light travels at a speed of 186 thousand miles per second. Thought energy transmission occurs one million times faster at 186 billion miles per second. Therefore Marcia's energy as a free soul in Heaven could easily reach the energy of my Divine soul intertwined with my body on Earth and also vice versa to allow instantaneous communication between us. Einstein's photons do not have mass and therefore do not have energy, but elementary particles do have mass and energy. In quantum theory, particles can communicate instantaneously with each other at great distances through energy transmission. Moreover God grants us His special Light along with a special Mass of His energy particles in the form of a Divine soul. Marcia could just as easily appear on earth in visible or invisible form although significantly more of her soul energy would be required to appear visibly transparent.

When Einstein spoke of energy and the speed of light, he was referring under ideal conditions to all forms of electromagnetic waves such as ultraviolet rays, x-rays, or electricity running through a wire, and not just the ones like ordinary light visible to our eyes. In the same way, we can feel the light of the rays of the sun upon us even though the sun is 93 million miles away and our eyeballs can lock into the majesty of stars projecting their brilliance billions of miles into the cosmos. Heaven we stated is one thousand times further than the sun 93 billion miles away. Yet, we can still be touched instantaneously by God's Divine Providence and feel His miraculous presence on earth. Under hypnotic regression I died in past lives and as my soul rose, it only took seconds for me to be in the spirit world. My soul travels at 186 billion miles per second but it took a little longer than the expected one-half a second to travel the 93 billion miles to the spirit world, because my soul paused and continually turned around to look at the dead bodies of David Samuels Smith and George Washington. Newton also observed this "seconds" transition to the spirit world from our earthly world with the eight-thousand plus clients that he treated.

Our Divine souls are energy, and as we climb the spiritual ladder and improve our character by making righteous moral choices, God can send us more energy of a higher potency so that our souls become even more powerful. The more of these potent energy particles and thus the more Mass we possess within us, the more intelligence and wisdom we acquire, since the energy particles that God sends us are mini brains incorporated into this Mass. Einstein is thus saying that mass and energy are different manifestations of the same thing. In life we make good and bad decisions leading to the possibility that the state of our energy is in flux. Because of the relationship of energy to mass and velocity, tiny amounts of energy particles added to supplement our souls can have a profound impact on our spiritual selves.

In all of our discussions thus far, we have been talking about the Divine soul living under the constraints of a human body. What about a soul in the spirit world free of the body? Is it more powerful? Indeed in Marcia's case, as you shall see, hers is. I can shut off the Devil's darkness and prevent him from talking to me, but I cannot turn off Marcia's energy. Light will always be more powerful than darkness.

Einstein predicted that space and time did not have independent existences but instead formed a fabric of space-time in a fourth dimension. Prior to the big bang, space and time also did not exist. According to biblical interpretations, God created the universe ex nihilo, out of nothing. However, as we are proposing, Light energy particles [God] existed and exist now, so can we really say there was nothing?

God had not yet released His Light and when he did, that's when the big bang took place. There was an expansion and a multiplication of God's energy particles that filled space. Along with the energy particles, there was a release of space entities and of helium and hydrogen from within God's Light energy particles that formed the gases of our universe's early atmosphere. The explanation would be that within God, all of our universe, including the darkness, existed. God, Who is Light, would then be a kind of reverse black hole. This latter possibility would be one way of assigning God as the Creator of all things. These Divine energy particles throughout our universe, the God Field, remain invisible and have not as yet been detected by scientists. Neither have we seen God although He may show Himself in the Messianic Age according to the biblical prophecies. After this book was initially published in mid April of 2011, a startling discovery was made of the Higgs boson particle and it's associated unseen Higgs field. We discuss this at the end of Chapter 12. We propose that the "Higgs field" and the "God Field" are one and the same.

After one of our sessions, Jules referred to a quote regarding a drop of water in the ocean. The quote stated "Everyone knows that a droplet of water exists in the ocean but only a handful of people know that the entire ocean is contained in that droplet of water." If you think of the moments before the big bang, God's Light resting energy particles were a compressed droplet of water which grew into an ocean, but He is still God, no matter how many energy particles of Light were compacted or He became with the expansion of our universe. Each energy particle is God. Analogously the ocean is really entirely in the droplet of water. God has not changed. Neither has the ocean. Water is still water.

Speaking of water, this molecule is essentially "alive" with electromagnetic energy. There have been electron microscopic studies by Japanese scientists where different waters were examined under powerful magnification. Pure, uncontaminated water formed beautiful colored crystals that could be seen under the microscope, while dirty, polluted water did not show crystals and took on an amorphous structure. When the pure water was talked to negatively, the crystals reverted to the amorphous structures, Similarly if plain water was talked to with loving phrases, the amorphous water was changed to the colored crystals.

Fascinating studies of the experiments with water have long been demonstrated in the negative and positive energy messages sent to our bodies. Negative energy causes us harm, while positive energy keeps the energy of our bodies in symphonic harmony. Negative energy or worse, evil energy, is detrimental to our health. In contrast positive energy is healing. At a macroscopic level, this is not difficult to understand. We talk to our pets nicely and even to our plants and they thrive. Everybody understands rejection brought about by negative energy. If this negative energy comes from evil people under the Devil's wing, then this could lead to death.

We need to next talk about this whole field of energy channels within our bodies.

8

❧

ENERGY CHANNELS

IT WAS BACK IN THE MID NINETIES THAT I HAD MY FIRST EXPERIENCE WITH ENERGY. I was bipolar, and in 1995 I experienced a vicious agitated clinical depression, where I was unable to sit still for more than thirty seconds. My only respite was the two hours of sleep that I got each night. My situation was so bad that I swallowed two hundred codeine and Tylenol pills in an attempt to kill myself. The agitated depression lasted for months and one day miraculously lifted. After my second suicide attempt, Marcia was ready to call it quits, but something or someone changed her mind and she stuck it out. I came out of the depression by a miracle. At the time of my agitated clinical depression, I was bankrupt of any spirituality. I was to find out sixteen years later who that Someone was who persuaded Marcia to stay.

Marcia was so distraught about how far I had plummeted in my life that she was willing to try anything. She located a Tai Chi teacher, whose name I cannot recall. I think his first name was Jim. Jim tried teaching me Qigong and asked me to feel the energy moving from one hand up my arm across my neck, down the other arm to my other hand. After several sessions I did feel something moving, but it wasn't until one of the last sessions that the energy sped up on its own and actually moved across space from one hand to the other. My hands were held apart chest level in front of me. I was stunned by the flow of electricity and the jet-like speed of the energy moving on its own in this circular tract in either direction. Unfortunately the Qigong, even with this amazing energy experience, did not alleviate the agitated depression, so I stopped the lessons.

When I officially retired to Florida in 2006, I sought out a Tai Chi instructor and found an excellent one in Gary Tong. I was sporadic in my attendance, but Gary did incorporate Qigong, also known as Chi Kung, into his lessons. After being with him for a time, I asked him if he could recommend some reading on energy channels in the body. Gary suggested I read *Taoist Cosmic Healing* by internationally renowned Qigong master Mantak Chia. I picked up the book and began my journey. I discovered that healing was more than energy channels, as sounds and colors of the organs play a role, along with an inner smile and proper breathing to set the stage for energy channel measurements. The author talks about so many aspects of energy or *chi* that it's not possible to discuss them all here.

I was particularly interested in the eight channels that Mantak Chia described. Later research by John and Matthew Thie, *Touch for Health,* seems to suggest that the body's surface has fourteen channels. The first was the Functional Channel (yin) that ran down the front of the body from the tongue and around the perineum and up the Governor Channel (yang) paralleling the spinal cord to the top of the head and down to the third eye in the center of the forehead. After I did the suggested preparatory work, I allowed myself to feel the energy moving at discrete energy points (*chakra* points) along these channels. I took some time at each energy point and then felt the continuous oblong movement along the front and back of my body, except at first I wasn't able to go from the third eye on my forehead to my tongue. You need to push your tongue up against the roof of your mouth as far back as possible. Finally after several unsuccessful tries, I was able to move the energy almost through my nostrils to my tongue and then down my chest. I felt almost like a jump in the energy to reach my tongue, like my earlier experiences with Jim.

I then allowed the energy to proceed along each leg, both the inner and outer skin regions, through the bottom of my foot and my toes, then back up the spine and to the outside and inside of each arm by way of the fingers. I didn't go any further, but I could easily have followed the other energy channels outlined in *Taoist Cosmic Healing*. From Gary I learned how to store this energy in one of the Tan Tien regions located just below our navel two inches inside of our bodies.

If you just do a standing pose called the Wu Chi posture and allow all tension to leave your body, you will begin to feel the *chi* or energy at your fingertips, while at the same you will seem rooted into the ground by the energy flowing through the soles of your feet. And then if you move your arms in the Tai Chi form, you will feel the energy moving with your motion. The same holds true when you hold an imaginary ball between your hands or pretend you are pulling toffee apart. As you become more expert, you can take in energy from the Earth below or from the Heavens above.

After a while I began to feel energy flowing everywhere in my body. I could just sit on the couch for hours and the energy flow never ended. It was moving on its own with no direction from me. Your blood travels miles through the body and so does your energy. There is no end and no beginning, so I believe had I wanted to, I could have gone on indefinitely.

What does this energy flow have to do with the energy equations described by Einstein? Energy travels at almost infinite speeds, so how can we feel it in our bodies? Our contention is that the energy is one and the same. The most plausible explanation is that akin to Einstein's theory of general relativity, the body acts as its own powerful "gravitational force" and dramatically constrains the energy, making it possible to track its direction and speed with the possibility of changing the speed from slow to fast or vice versa. I can alter the speed of my energy in my body.

Einstein would have loved this explanation because it refutes the uncertainty argument of quantum mechanics where you cannot determine the momentum and the precise location direction of a particle like a moving electron simultaneously. Einstein could not conceive of probabilities and randomness in God's nature, while quantum mechanics had moved on and was already looking at predicting the future as a series of probabilities.

The body is kind of acting like a black hole, slowing the energy speed to an almost complete stop. As Isaacson points out, in black holes space and time lose their individuality and merge together in a sharply curved four-dimensional structure precisely delineated by Einstein's equations. Even the light energy of our sun could be swallowed up by a black hole's gravitational pull. In the spirit world, the free soul is travelling at warp speeds, and time and space have no meaning, just as God is independent of time and space. There are no black holes, only the pure light of the soul world that Dr. Michael Newton's clients repeatedly experienced in their past lives-spirit world regressions.

Telepathy is possible when open communication is required from the spirit world to the physical world, Marcia's Divine soul to my Divine soul. Reverse communication from my soul to hers, however, requires my energy to move at similar warp speeds permitting instantaneous communication between us. We shall demonstrate that this is possible during "thought" which originates in our Divine soul and makes energy telepathy possible. No one has satisfactorily described the origin and nature of a "thought." No part of the human brain has been delineated as totally responsible for what we term "thoughts." We don't know how thoughts come into and leave our brains.

Until now I hadn't looked at Mantak Chia's book for perhaps five years, and I was pleasantly surprised to find a card with a cardinal's picture on the front stuck within the pages. Marcia and I had put up a bird feeder and we used a birdseed that we hoped would attract cardinals. Sure enough, a husband and wife cardinal came to enjoy the pickings, along with bluebirds, blackbirds, and a woodpecker. The squirrels too had their stomachs satisfied until we realized that they were hogging the feeder. We switched to a squirrel-proof feeder and once again the birds were able to eat.

I don't know what this means, if anything at all, but as I'm writing these words, I am also turning my head to the television to watch the ball drop in Times Square in New York City, bringing in the new year, 2012. I feel totally lost in my life without Marcia's physical presence, so I'd like to quote from Marcia's just-discovered cardinal card. Don't get me wrong, I feel blessed to have her spirit soul with me. It's just not the same as holding her and kissing her. I miss her so.

My Darling:

You are a wonderful father and the best husband I could ever imagine. I love and respect you very much; in your calm steady way your inner strength always shines through. All of my hopes and dreams center around you. I'm so grateful to share my life, my love, and our home with you.

I'll love you forever and look forward to many wonderful times in the coming year.

Yours forever,
M.
p.s. Our cardinal, the newest addition to our property.

When we returned from India, Marcia had severe diarrhoea problems. She was tested for parasites on two occasions, but nothing ever showed up. An infectious disease doctor couldn't come up with a diagnosis. Marcia wanted no part of doctors and desired to pursue a holistic approach.

I found a clinic in Delray Beach, and we began several years of homeopathy, until the doctor in charge of the clinic was institutionalized with dementia. Oddly enough the approach involved homeopathy, but it was a different kind of treatment based upon energy transfer of homeopathic agents through space into three dropper bottles containing distilled water; or in one bottle, several drops of Marcia's blood were placed into the dropper bottle of distilled water. All this was done by a special machine manufactured in Germany.

To transfer energy from these homeopathic agents in air seemed like hocus-pocus to me. However, as Marcia began to show improvement, I decided that I too wanted to try this mysterious technique to improve my immune function. The methodology strangely also worked on improving environmental and food allergies. There is no doubt that the treatment reduced the diarrhoea and by supplementing with good bacteria probiotics, Marcia was almost cured of her diarrhoea prior to her diagnosis with liver cancer.

Coincident with the intestinal symptoms was a marked weight loss, which originally I thought was due to the diarrhoea. However, I remember remarking to Marcia that the drop in her weight was too dramatic. Her answer at the time was that she wanted to be at the weight she was when she was twenty. In retrospect the way she looked at me, I realize now she was not telling me the truth. Indeed the weight loss was a separate issue due to the progression of the cancer. She just confirmed this with me now in telepathy. She lied to me because she wanted no part of doctors, even though she knew many years ago about her cancer. We are getting ahead of our story, so we shall come back to all of this shortly.

After my first past life-spirit world regression, I spoke to Jules of scheduling a session on energy measurements. I was to learn that when you pose questions on health or good and evil or the soul, you are measuring energy in a distinct way that is absolutely credible as it was repeatedly accurate. Marcia and I have concluded that we are measuring the energy of the soul by Jules's methods. Jules, however, was not aware that this was the energy of the soul that started out in the foetus as being all good because the soul is a gift to humans from God. The Divine soul entering the foetus from the spirit world has been healed according to Newton's thousands of patient studies and thus begins life one hundred pure percent and positive. As we age, the soul encompasses all the positive plus other people's negative energy running through all of your organs and cells in addition to the surface energy channels described by Mantak Chia and the Thie(s).

I learned from Jules that you can train yourself to measure the soul energy inside your body. Illness is not a cause of negative energy but is likely in some instances the result of negative and/ or weak, stagnant, blocked energy. However in evil people where darkness has covered the great majority of the Light of their Divine soul, analogous to a partial eclipse of the sun, the energy they feel and possess is mostly dark negative energy. Marcia and I have determined that if their Light energy drops below 18 percent, these dark souls will die. Evil people can intentionally both cause and exacerbate illness within you. In the Devil's case, God's Light is covered over the entire 100 percent. The Devil has both human and part angelic attributes so God has chosen to keep him alive even without any Light energy particles. The Devil is the fallen Angel of Death and has been alive these millions and perhaps billions of years.

One more very important point worthy of note. God can convert negative energy within you to positive energy. He can also and will destroy evil people by dropping them below the 18 percent mark. In biblical days, the non-Jewish prophet Balaam was ordered by his king to curse the ancient Israelites with evil energy as they passed through the kings land. However all of Balaam's curses were turned into blessings by God

Asking questions of your Divine soul allows you to measure the positive and negative energy as well as evil dark energy in yourself and other individuals in very specific ways so that you can answer the questions you pose. What we are measuring in terms of body-soul positive and negative energy taught to me by Jules in the Yuen technique is different from the external body energy channel measurements. The energy circulating in the eight energy channels described by Mantak Chia or the more recent work by the Thie(s) suggesting fourteen surface energy channels, or Yin and Yang meridians, is magnetic so that the energy expressed in the body is like a magnet with a negative and positive pole. As the Thie(s) discuss, "along the meridians there are approximately five-hundred electromagnetic acupuncture points that consist of small palpable spots that can be located by hand, with micro-electric voltage meters, and with muscle testing. The skilled chiropractor or practitioner through unlocking and strengthening weak muscles releases blocked energy in the meridians. The same result is accomplished by the acupuncturist using tiny needles inserted into the acupuncture points in clever ways to bring about specific improvements in energy flow and optimization of human performance."

Let's look closer at soul energy to see the difference.

9

SOUL ENERGY

A T FOUR YEARS OLD, I WAS ALREADY SUFFERING WITH HEADACHES. My mother fed me codeine and aspirin pills, which were and still are available without a prescription in Toronto, where I was born. For the next twenty years, I took tens of thousands of these pills to try and alleviate the pain. After completion of my PhD in biophysics at the Weizmann Institute of Science in Israel in 1969, I returned to Toronto. Something happened in this geographical transition, because the muscle tension headaches that I had suffered with since childhood turned into incapacitating migraines. The codeine and aspirin no longer worked, although I kept shoving them into my system.

When I met Marcia in 1981 on Long Island while I was working at Stony Brook University and in the middle of a separation from my first wife, still no relief was in sight from the physical migraine pain. For about the next seven years, I used Caffergot suppositories, and if I was lucky and took the medication at just the right cycle of the dilation phase of the pain, I could shorten the intensity and duration of the migraine. This went on until I discovered in primal therapy, which I began in 1981, that the migraines were caused by repressed anger and rage. When I went back in time to my mother's womb and felt this rage, the horrible constriction and dilation phases of the migraines disappeared, but I was still left with a strong residue of pain in my neck, head, and shoulders. At least the pain was tolerable. I remember vividly Marcia suffering alongside me in one of my most severe migraine episodes, although they were all bad and occurred on a weekly basis..

When we made the next geographical move to Florida in retirement in 2005, a new form of strong pain came that affected me on both temples. It was the latter pain that I presented to Jules after our first session on past life-spirit world regression. "Jules, is there anything that can help?" "Yes," he responded, "The use of energy to unblock the pain." I was hopeful for the first time in years that maybe I could be free of the pain. It hasn't exactly worked out that way, as the temple pain continues despite my increasing knowledge of energy measurements and transmission. I now believe that the pain is due to a multitude of food and environmental allergies. Somehow the physiology of my body shifted when we moved to Florida as it had done when I came back to Toronto from Israel.

Jules casually described his own personal experiences whereby he was on a phone call with a friend in California who cured him of his pain. His story fascinated me. How could this be? And how was it done so quickly? Over many years Jules has pursued different ways to keep the body healthy, and he has been to many workshops and has studied with various teachers. I asked him if he could teach me and he said he would. He gave me reading materials and two weeks later we were talking energy. Jules never realized that what he was teaching me was the energy of the Divine soul. Neither did I at the time. It took me several months to put the pieces together and to perfect my own Divine soul energy measurements. If you venture into this world of energy, you must be absolutely sure not to introduce your own bias or misperception as you need to eliminate negative or positive "thought." If you do think and are not absolutely objective, you won't get the right answer. When my daughter was down for a visit and I taught her some of the techniques, we wound up getting different answers to the questions we posed on good and evil. I couldn't understand why until I realized that we were both experiencing outside interference from the Devil. For the first several months, it was difficult to block this evil interference, but it is no longer a problem as I have perfected my measuring skills. In fact, I don't even have to forcefully measure anymore. I just command my energy to seek the truth in the absence of bias or the Devil. My energy does its own thing involuntarily. In fact, my entire body acts upon command. If I say dance the waltz, I begin doing the waltz. If I say move my neck randomly, it moves involuntarily in all directions. Just what is this soul energy?

June 16, 2011

Jules started out by telling me that he used muscle testing and internal energy measurements to correct for abnormalities of the entire anatomy of the human body. He knew the muscles and nerves that needed to be corrected, and the corrections were done before you could blink your eye. The techniques were multipronged and focused on emotional and physical well-being. One of the books he recommended was Dr. Bradley Nelson's *Emotion Code*. Dr. Nelson suggested that we could turn the energy on and off in our bodies, which I have learned to do at will. He used muscle testing to determine whether the energy is weak or strong within us and was able to teach his patients how to feel energy flow and feel the areas that are blocked. He then used magnets to disrupt the electromagnetic energy in either the functioning or governing energy channel and release trapped emotions. I know this simple technique using even refrigerator magnets to be true from my own experiences and from helping Melanie. Jules emphasized strongly that I should not trust anyone, even my own mind and its thoughts and ideas, because he stressed that I needed to feel the energy for myself, as this allows me to distinguish truth from lying in another person.

His first question to me or rather my soul was, "Is your name Jerry Pollock?" I looked at him curiously and before I could answer, he asked me if I felt any motion inside my body. Jules wasn't aware of my past energy experiences in Qigong, and to tell you the truth, I never mentioned them because I was so focused in the moment on finding Marcia in the spirit world. I have been a student all my life, and I was captivated by what I was learning from Jules.

I responded to the question, "Yes, I feel the energy around my heart." He was pleased and commented that he liked working with me because everything he taught me, I immediately picked up.

He then invoked a No response by asking me something like, "Are you Jackson Pollock the famous artist?" I shook my head, but at the same time felt the energy drop down. "I feel it below my navel."

Jules replied, "That's a No answer where you felt the energy." We went on like this for some time so I could get the hang of it, but I was getting some questions wrong.

Jules then said, "Imagine a colored light going from your shoulders to your genitals inside an empty cavity in your body." He stated that it would help if I could imagine space for example in my brain or heart or fingernails for our next regression session, as doing this allows you to more easily sink into Newton's state of super-consciousness, required to flip into the spirit world from a past life regression.

"I can do this," I said. I asked. "Does it matter what color the light is?"

Jules uttered, "No, the light can be white, yellow, gold, blue, whatever you want." I have used all these colors, but I am partial to blue. I also imagine this light to be a sword moving in the space within my body from the neck down to my groin. Jules instructed, "As you place the light inside yourself and visualize it, say 'Correct neutral' for misperceptions when you are answering a posed question."

The founder of this technique was a man named Yuen. Jules said that the technique had broad application. For example, one could first determine through heart-navel energy measurements within yourself if you possessed negative energy passed from another person. "Do I have negative energy?" A Yes answer is felt around the heart, while a No answer is felt below the navel. You can rephrase and ask the question in different ways. "How much negative energy do I have in absolute percentages?" Your negativity would indicate 100 percent negative for a Yes around the heart and a No answer below the navel would be 0 percent negative. If your measurements of energy are detected in between the heart and the navel, then the percentage negative energy will be somewhere in between. You can therefore refine this even further to determine the exact amount of negative energy you or another individual has if you ask the question as a scale of percentages from 0 to 100 percent. If you then place the light through yourself, you can correct neutral and eliminate whatever negative energy percentage you have from another person at any one moment in time within yourself. When you ask the question, "Do I have any vestige of negative energy inside my body as of now after my correction?" you should feel the energy below your navel indicating No, or there is no negative energy within me.

Jules then explained how you can remove passed negativity within another individual by placing the light through the other person. You don't even have to know or see that other person. Before you do this, Jules stressed you should ask the question, "Do I have permission to treat the person or remove negative energy?" If you do get a Yes, then you need to determine how weak or strong that person is to you. The optimal situation is where you place the light through the other person and make that person strong to you and then place the light through yourself and make yourself strong to the other person. Once that's done, you can help the other person. Since negative energy inside you indicates mal health, the idea is to remove this negative energy and restore health. In this same way, you can measure trapped emotions. You can also do this by muscle testing. Removing the trapped emotions, however, requires the use of a magnet to disrupt the energy in your body.

When Jules, who is very experienced at all of this, measures energy, he feels the energy immediately within himself at the exact moment he is measuring the energy through muscle testing. Both the muscle testing and the Yuen technique should give you the same answer if you are totally objective and eliminate your own bias and evil interference. As we discussed, Jules referred to this energy simply as energy, but Marcia and I have determined this to be the energy of the soul that has spread from its initial location in the brain to all sites throughout the body.

At this point in time in the training, I was on my own in measuring the positive and negative energy as it relates to health, but then Jules introduced concepts of good and evil. He said, "People can be entities or ghosts, which are souls that have left a dead body and never returned to the spirit world." Newton also discusses such souls but says that they are eventually recalled to the spirit world for rejuvenation. Such souls of ghosts may remain in the spirit world and not have another incarnation for a thousand years, while a normal soul may incarnate into a new body as quickly as thirty to fifty years or even less according to Ian Stevenson's studies with young children.

Jules also spoke about demons as souls that never had a body and are creating havoc toward people still alive. Marcia and I classify demons as evil people, those whose negative dark energy in their Divine soul usually hovers around 80 percent having barely enough Light energy to be alive. Jules refers to my definition of demons as scumbags. There are also people whom we classify as associated demons who have a demon around them and who can be flipped to become a demon.

Jules then discussed curses and spells. He said everywhere he went, he would clear the area, inside or outside, of entities, demons, curses, and spells. Say you are in a restaurant and you want to clear the restaurant of negative energy. Simply allow the light to move from floor to ceiling or wall to wall, and you can clear the room of negativity. You can also do this for your car and your home or for someone else's home and car. The sane holds true for clearing an animal of negativity. My energy has been refined to where I can rapidly flood my entire house with light moving rapidly from room to room to eliminate negative energy.

Jules discussed the source of negativity. He said that it comes from a person possessing evil energy, such as a demon or associate demon but non-evil negative energy can also be transmitted. The negative or evil person can hug you and pass on his negative energy, or he can send you a birthday card expressing well wishes written out from his pen or he may even transmit negative energy through his voice on the telephone. He may have touched a photograph he has given you or sent you a gift that he has touched, or he may have the ability to cast a curse or spell on you. Witchcraft is real just as the powerful Voodo from Haiti is. But you can remove all of this passed negative and evil energy by imagining a shining light through yourself. And you can muscle test or energy test to make sure you have rid yourself all of the negative energy. We ended the session.

I could barely concentrate on my driving on the way home. Questions filled my brain which wanted to burst like a cornucopia or a stuffed piñata that Mexican children smashed with their sticks. Who was I talking to when I ask permission to remove negative energy from someone? When I checked my internal energy, the answer came up that my request was to the other person's Divine soul which could answer Yes or No. If the individual's soul was 80 percent darkened, the No answer always came in the form of the "f" word, as I was sworn at with expletives when I made a request to remove his or her evil negative energy.

Bradley Nelson in his book, *The Emotion Code*, uses magnets to disrupt the normal electromagnetic energy channels in a person's body to remove trapped negative emotions. The chiropractor uses muscle manipulation at specific energy points and the acupuncturist disrupts and improves the energy flow by needles inserted into the acupuncture points along the meridian channels.

By the same token, a Tai Chi-Qigong master can remove both negative and blocked energy and heal you by passing his positive soul energy into your body without touching you. I've witnessed the latter with real live subjects in a holistic conference. At the meeting, the Qigong healer stood at arms length standing behind the subjects sitting in chairs. A word of caution! The healer could be passing on both positive and negative energy to you. It would help if the healer has the ability to take in more positive energy from the Earth and Heaven in order to ensure a higher percentage of positive energy and minimize the transfer of negative energy. We mentioned earlier that energy transmission between particles travels simultaneously in both directions. Therefore each time the healer does his energy transfer, whether at short distances like at the conference I attended or with healing across the globe at long distances, the healer is picking up negative energy of the people he is trying to heal. And you are actually receiving a mixture of both positive and negative energy unless the healer has the means to filter out the negative energy before passing it onto you.

The placing of ordinary light electromagnetic energy within you or within another person in the Yuen technique specifically removes unintentional or intentional negative and evil energy passed to you or them from another person. It also is measuring the energy of your Divine soul so that you can ask questions about truth and lying, good and bad, and the function of your soul. If this negative energy passed to you from another person is not removed, it can lead to ill health and for example the trapped negative emotions described by Nelson. You then may require the services of a chiropractor or acupuncturist or Qigong master to improve your health. If evil people intentionally pass you their evil energy, commonly referred to for centuries as the evil eye, death can eventually result.

As we describe in Chapter 11, the Yuen technique doesn't change the moral choices that you make which if bad can cause a darkening of your Animal soul receptors. Increasing and lightning up the receptors of your Animal soul can only be achieved either by improving your character or by Divine Providence. If you believe in angels as I do, then it is not a stretch of my imagination or my faith to accept that Raphael, the Angel of Healing, can heal you by passing God's special positive Light energy into you. This is different than simply humans disrupting or removing negative blocked energy to improve your health and is infinitely greater than the positive soul energy of the Qigong healer. God Himself has done this dramatically in the bible prolonging king Hezekia's life by fifteen years when Hezekia was certain to die and was told so in no uncertain terms by the prophet Isaiah. I believe God lifted my hopelessly severe agitated clinical depression back in the summer of 1995 at Marcia's request. Much more than that, He has made my Divine soul whole and pure by leading me in His direction of spirituality. Back in 1995 I didn't have the unshakable faith that I have today.

When my daughter Melanie came down from Long Island to visit me in Florida, I taught her these techniques and began to measure our family and friends for negative energy brought on by other people. Negative energy is different than evil energy in that its source is not from evil people but from generally pessimistic, materialistic, and non-spiritual individuals. After making our family and friends strong to each other by optimizing energy connections and also to potential evil influences from undesirable individuals, we were amazed how much negative energy we found. We then shone light through our family and friends by passing light through them, but it was often necessary to repeat this if they were exposed repeatedly, which they were.

The fear of evil energy in many cultures has been with us for centuries. As stated above you may have heard about the evil eye and to stay away from people who have it. To protect themselves, people have worn religious and mystical amulets or the red string of kabbalah. Before I learned these techniques, I routinely purchased and wore the red string with the Jewish Matriarch Rachel's energy that Marcia tied around my wrist. Every year the Kabbalah Center in the United States travels to Rachel's tomb near Bethlehem with Israeli soldiers and wraps red yarn (red string) around the grave to acquire Rachel's energy. Now in place of the red string, I use the Yuen technique. However, I probably should return to wearing the red string, and I have, as a preventative measure against passage of negative energy from evil people.

I only did the three sessions with Jules, but after my last session Marcia's image appeared before me for the next couple of days. Even before I saw Jules, something that I thought was very strange happened. When I went close up to Marcia's enlarged 18 x 24 inch picture, which I had framed, the eyes in the picture moved. Of course my family members could not see this strange happening, nor could they see that Marcia in the picture moved both her head and shoulders and followed me around the room with her eyes directly upon me. I even went outside the house and looked at the picture at a sharp angle through the window, and Marcia in the picture moved her head and shoulders in the opposite direction looking out through the inside of the window at me outside. She even anticipated my walking outside as she was already in the correct pose when I arrived at the window. She couldn't see through walls but she knew that I was thinking that I would go outside to view her picture. Again I tried to get family members to see what was happening, but again they thought something was wrong with me. They worried that I was returning to my bipolar disorder state and was psychotic and hallucinating. I wondered that myself and told Jules about what was occurring. In addition someone was moving pictures around on our family room table where we have framed family photos. One of these framed pictures was even tipped on its side. It remained that way for about a week, and then it was flipped right side up. Then after a few days, the picture was placed on its side again, only to be flipped back right side up again.

I did some research and discovered that a soul could do this with pictures, and I wondered if this was Marcia. I invited Jules out to dinner a couple of months later after the sessions as an acknowledgment of my appreciation, and I told him what was happening in the house with the pictures. I asked him if there was some way to contact Marcia in the spirit world and right then and there in the restaurant, we made Marcia strong to me and me strong to Marcia by putting a light through myself and through Marcia's transparent soul as I saw her in my past life-spirit world regression. At dinner Jules also taught me some new techniques and we had a delightful evening.

After we left the restaurant, we stood by my car talking. Jules asked me to look into the distance but not to focus on any one object. The idea was to see an object within a series of objects without narrowly focusing on the object. I remember looking in the distance at a building, but I was also looking at the same time at everything around the building like trees, a lamppost, and another building. A strange thing happened at that point of avoiding narrow focusing. I found myself involuntarily doing a 180-degree turn facing the restaurant from which we had just finished dinner. However, I wasn't turning my body. The energy in my body was turning me. I blurted out, "Jules, there is a demon in there."

I now find that if I open focus my vision, my head will be turned by my energy if an evil person is in my presence. This can happen when I'm driving and if I'm passing a house full of negative energy. Jules had recommended reading *The Open-Focus Brain* by Les Fehmi, which explains the phenomenon of avoiding narrowly focusing on objects. By doing this we apparently take ourselves out of stressful beta waves and place ourselves in a more relaxed alpha wave state.

I went home and over time I began to gains snippets of conversation from Marcia. Each time I heard her voice, I checked my energy levels to be sure it was her. At the same time I was scared to death to go to bed, because each night I was being visited by ghosts and spirits who appeared in the dark when I woke up during the night. My sleep grew more and more restless and my dreams became more and more like anxiety dreams. I started sleeping with the light on, and although the ghosts and spirits more or less went away, the restless anxiety dreams did not.

As I continued to talk to Marcia's soul, I noticed interference from a male voice who introduced himself as the Devil. Now I was really worried that I had returned to mental illness. The Devil was continually swearing at me and I was swearing back at him. He wanted to own my soul but I told him where to go, that I was not for sale. After a while I learned how to prevent him from talking to me, and for the most part now I talk to him when I wish to and not when he wants to. Our encounters made me less and less afraid until I felt confident that I could handle my fear and handle him.

I once read that the Devil is grateful for someone who will stand him down. I also knew from my biblical research that the Devil was a creation of God and not an independent entity as the Devil believes he is. In fact my not-so-fictional novel *Gog and Magog: The Devil's Descendants* seems to be coming true as far as the Devil being alive is concerned. I continued to talk to Marcia but there was a new twist. The Devil could speak exactly like her, so now I had to check each time to confirm whom I was talking to, Marcia or the Devil. I can also correct to remove the Devil from the conversation but I have to be vigilant about it.

As time passed and the fall season came, friends visited me at the house and to my surprise could now see the picture of Marcia following them around the family room. My nephews and nieces came down from Orlando and stayed with me, and they too saw the head and shoulders of the picture follow them around the room. Even Marcia's brother Les, who was visiting, saw the picture move and look at him. It was particularly important in Les's case as he is a sceptic. Everyone who had the experience was shocked, and a couple of friends even saw more white in the eyes as the eyes moved up and down or round and round in a roll.

Thank God I had been through bipolar disorder in the nineties. I had heard voices in my head, but the difference was that in my psychosis I wasn't able to turn these voices off until I took anti-psychotic medication. Here I could turn off the Devil's voice at will and I felt perfectly normal. I wasn't crazy, but what was the explanation for Marcia's picture motion?

I measured my energy very carefully and determined that the energy of Marcia's soul in the 18 x 24 inch framed picture represented 2 percent of her total energy. The remaining 98 percent was in Heaven in the spirit world. No matter how many times I checked it, the explanation was that Marcia's soul energy was in the picture and was moving her head and shoulders. I also made 8 x 10 inch copies and Marcia moved in these pictures. Since these smaller pictures too represented her soul energy, I wondered if even more of her spirit world energy is required. It turns out that my energy measurements suggest that it is not. A much smaller percentage of her soul energy, on the order of one-hundred times less per picture, is needed because this smaller amount of energy is coming from the "master" 18 x 24 picture. Marcia can move and remove her energy at will and if she doesn't want certain individuals to see the picture move, they won't see it. This is what happened initially with my family members who didn't have an open mind. Later, some of these same family members did see the picture of Marcia move. If I got real close to the picture and looked at my *luvee*, it was eerie as the subtle movements made it seem like Marcia was alive. I could even ask the picture questions and if I carefully observed her right eye, it would move up and down for a Yes answer and roll in a circular motion for a No answer. In thinking back about my first date with Marcia in 1981, there may have been an unconscious recognition when our eyes first made contact. The eyes are the window of the soul, and our souls may have recognized each other from past lives that we shared together although our bodies did not.

Then one night in early December, I was talking to my nephew Scott when his Aunt Marcia, my Marcia, interrupted our conversation and was making suggestions through me to him, as she had always done in direct conversations with Scott when she was alive. Marcia was the most important person in Scott's life. She was his trusted confidant and he was extremely sad about losing her.

I hung up with Scott and from that moment on Marcia started to freely converse with me. I was in awe and asked her if she was talking to me from the soul world. She confirmed her soul energy in the picture and said that she was fulfilling her commitment to be with me forever. The Creator had given me another gift. A few days went by when we pledged our love to each other when I realized that she could read my thoughts. When I asked her about it, she confirmed that she indeed knew my thoughts. I wasn't worried because two months after she died in March 2011, I went up to her picture and told her that we were soul mates and we should share all of each other's thoughts. Then and there I began to share everything on my mind, good or bad, as I felt soul mates should. During our marriage we were so much in love, but we did not do this. Now I wanted to start off totally transparent. At that time her picture had just begun to move.

There was one thing she would not do. She would not and still does not interfere in human affairs, even though she knows everything that is going on in the physical world and even foresees the future. When I asked her if all souls in the spirit world could read minds, she said No. Then in mid-December, I got a brainstorm. I asked her if she would co-author a book with me because of her knowledge as a wise free soul who is close to God. She did not hesitate, to my surprise, and agreed. I suggested the title, *Putting God into Einstein's Equations: Energy of the Soul*, and she loved it. I wanted her name to appear first, but she insisted that mine be before hers.

Everything that has been written prior to now has therefore been confirmed across two worlds by my *luvee*, from the soul of Marcia in the spirit world. We are not just espousing wild theories, although we realize that we are not supporting any of our hypothetical statements by traditional scientific double-blind experiments. We communicate by thought where the energy is transmitted telepathically between our Divine souls at one million times the speed of light.

Not so long ago, I did a simple experiment to determine whether energy follows thought or thought follows energy. I took a deck of playing cards and turned over the four of spades. I then thought about the question, "Is this the four of spades?" and when I measured the energy, blocking bias and evil interference, my energy immediately traveled to my heart, indicating a Yes answer to the question. Actually, it didn't matter whether I thought the query silently or said it as a thought aloud. I still got a Yes answer. If I asked, "Is this the six of spades?" I got a No answer as expected with the energy moving counter clockwise below my navel.

Then I thought to myself silently, "This is not the four of spades. It's the six of spades." When I again asked the question, "Is this the four of spades?" while looking directly at the four of spades, my energy immediately dropped below my navel, giving me a No answer. I was looking directly at the four of spades and thinking it was the six of spades, and my energy told me that the card was not the four of spades even though it was. I concluded that "energy follows thought" and the transition is immediate.

I asked one more question in sequence to the above questions. I was looking at the four of spades, but I was thinking it was the six of spades. I thought, "Is this the six of spades and not the four of spades?" As expected, energy followed thought and not reality, and I got a Yes answer as the energy moved from below the navel back up to the heart.

I have now just measured my energy asking the question, "Does a thought originate in the Divine soul?" The answer I got was a Yes. Marcia has just confirmed this through measurements of her soul energy, which is much more powerful than mine. My soul is stationed within the constraints of my physical body. Actually her soul energy as a free soul in the spirit world is about one hundred times more potent than mine inside my body. Therefore she can read my thoughts and finish my sentences, but I cannot do likewise with her.

If you ask the question, "Does the human brain originate the thousands of thoughts we have on average each day?" The answer is No. Even before I could ask the question, Marcia had read my thoughts and answered No. Next question, "Does the energy get transmitted telepathically at warp speeds from my Divine soul located inside my brain to Marcia's soul in the spirit world?" The answer is Yes, and again Marcia answered Yes before I could even ask the question. I think she should be the first author of this book. Before I could get this statement out, Marcia's soul answered No.

"Does the energy get transmitted telepathically only after a thought?" Marcia seemed to be testing me because she was not answering me. Either that or she was not answering because the Devil was lurking about. Measuring my energy and getting rid of the interference from the Devil and being totally objective, I got a Yes answer. Marcia then chimed in a Yes and said that the Devil was trying to interfere. She therefore knows when the Devil is about, but my soul energy cannot as yet detect him, again showing the potency of her soul versus mine. I know, however, that I'm getting closer in doing this. Some biblical rationale does exist for God inserting angelic energy into individuals when they need it.

"Can the energy circulating in the body transmit on its own in the absence of thought?" The answer is No, as confirmed by Marcia. "Can the energy circulating in the body transmit telepathically at one million times the speed of light if there is a thought?" The answer is Yes, because the soul energy in the brain is the same as the energy circulating in the body. "Is Marcia connecting to my total soul energy?" Again a confirmed Yes answer.

According to Michael Newton, the Divine soul incarnates initially into the brain between three months after conception and birth. After integrating with the foetus's neurological circuits, the soul and body try to act together to provide harmony to a new life. "When does the energy in the brain spread to the rest of the body?" It happens simultaneously when the baby is born, as confirmed by Marcia. The energy of the soul from the brain spreads to all body organs and cells. "Is each soul energy unique?" Our answer is Yes, partly because we have led different past lives and partly because God assigns us unique pieces of His energy prior to our incubating inside our mothers' womb in a new incarnated foetus.

"Do we carry the soul energies of our past lives, and is this how we are able to recognize and proceed under hypnosis to past life regressions?" The answer is Yes. "So does each person receive a new piece of God's soul in addition to all the soul energies from past life regressions so that their Divine souls are a composite of all the incarnations over time?" Another Yes. "Do past life bodies play any role in shaping incarnation into a new body?" The confirmed answer is partially. I thought this answer was strange. When I asked Marcia, "What percentage of each past life soul exists in a soul incarnating into a new body?" I got a surprise answer: "It's only one percent, Jerry."

I then asked, "Where is memory recorded?" Marcia answered that both the human brain and the soul recorded memory as life was being lived in the present. In most souls constrained by the body, there is no recollection of previous past lives but as Tom Shroder describes in his book, *Old Souls*, based upon Dr. Ian Stevenson's work, there are young children who vividly remember living a very recent past life without going through hypnotic regression.

"Does God determine who gets what?" Yes and No. "Does God select soul energies for certain individuals and are the rest assigned randomly?" The confirmed answer is Yes. God selects certain individuals in new bodies initially based upon their spiritual potential. Do we all start out with equal energy potential?" The answer is Yes. "What determines the future potency, both in a qualitative and quantitative sense, of one's soul energy? Is it leading a righteous life?" My energy moved between the navel and the heart, suggesting that the answer to the question is partly Yes and partly No, which Marcia confirmed. "Does leading a total spiritual life including being righteous generate maximum soul energy potency and a partial spiritual life less potency? Can energy potency change over the course of one's life?" The answer is Yes on both counts. There can be less or more potency depending upon one's moral choices and spiritual behavior in life. "Does the soul energy potency determine who will enter the Messianic Age irrespective of a person's religion or beliefs?" This answer is important. It is Yes.

There was one aspect of Marcia relative to me that was still confusing for me. When I recently was watching a movie in the bedroom, she said she was watching with me. She told me she was communicating with me telepathically, but she wasn't with me in the bedroom. What was with me in the bedroom was about 2 percent of her energy that she sent down to the bedroom to watch the movie with me. Although the energy in her picture in our family room remains permanently there, the energy she sent down to me in the bedroom, which was invisible to me, could be retrieved after the movie was over and I went to sleep. Marcia had the power to do this at will.

The reason I couldn't see her is that a larger percentage of her soul energy is required to create her transparent human form. Marcia just told me that it varies for different souls, but for her to do this requires about 15 percent of her total energy, I asked her, "How much is required for telepathic communication between us?"

"Smaller," she answered. "About 5 percent is needed from both your soul and mine."

To modify the butterfly at her gravesite, making it unique with blue and red circles on its wings, Marcia needed to use about 15 percent of her energy, which is recoverable at any time. Another point to be made is that just like our soul energy can increase, so can the energy of the soul in the spirit world. Deceased souls do good deeds on Earth and help loved ones both on Earth and in Heaven. Therefore if Marcia's energy is expanding for good deeds and is also expanding because our universe is continually expanding, then the 100 percent has to be calculated upwards. Marcia's rate of increase is far greater than mine, so there is no telling how powerful she will be in the future. As we shall discuss in later chapters, God promises even more soul energy for us in Messianic times.

It is important for the reader to note that Michael Newton pointed out that even if the soul divides like that described for Marcia above, you need to consider each percentage of the soul as having all the attributes of the entire soul. You cannot divide the soul energy just like you can't separate the droplet of water in the ocean from the ocean. The soul is a soul. Such teachings are in agreement with the interpretations of Judaism and other religions.

Marcia and I could go on and on with a long list of queries in question format, but instead we ask the reader to assume that in the narrative to follow, Marcia and I have asked all the questions and confirmed them between us. Before we proceed to more aspects of the Divine soul and ask questions about God and good and evil, we should ask the question, "What is spirituality and how do we climb the spiritual ladder?" Since the Messianic Age implies eternal life, this is the basis for our striving to lead a meaningful life while we enjoy God's gift of our material world. Marcia always said, "Life is a gift." We don't believe that we need to spiritualize our material world. We believe that both worlds are wonderful on their own and should be enjoyed. A wise sage once figuratively said, "Hang onto Heaven but keep both feet on the ground." Marcia also said, "Everything should be a balance in your life as it was intended by God for nature."

We know that Divine souls in the spirit world unencumbered by the body are spiritual. How do we get closer to God in our physical world?

10

⁓⚬⁓

THE SPIRITUAL LADDER

OME YEARS AGO WHEN I WAS DOING THE RESEARCH FOR MY SPIRITUAL MEMOIR, *Divinely Inspired: Spiritual Awakening Of A Soul*, I CAME ACROSS QUOTES FROM BIBLICAL COMMENTATORS THAT MADE AN INDELIBLE IMPRESSION ON ME. Back in the 1990s when I had no spiritual resources and faced bipolar disorder at age fifty, I swallowed two hundred pills and came within a hairsbreadth of departing this Earth. If not for a Divine miracle of God coming to my wife, Marcia, I would not be here today. You can read about it in *Divinely Inspired*—that is, if you are so inclined. I summarize one of the quotes: "*In the time of extreme crisis, a person becomes incapable of calm reasoned analysis. His greatest strength at such a moment is the instinct he has developed through all his years of living and striving. It is too late when the awful moment comes to make the preparations or develop the personality to cope with it. People who have failed to develop their spiritual resources before extreme crises strike will not have the resources to conquer it.*"

In the world we live in today, a lot more extreme crises are taking place because of the economy. People are losing their homes, jobs, and marriages, stealing, and falling into clinical depression. The stress factor is enormous, and I wouldn't be surprised if we see increases in the rates of deadly diseases such as cancer, which itself can be classified as an extreme crisis. Suicide may be just waiting in the wings. Another biblical sage offered the following quote, *"In the modern world, we are intimidated by sounds and messages. Some we ignore, others we hear, a few we assimilate, but only a rare one in a lifetime changes us."* My hope is that this message helps you begin or continue your journey on the spiritual ladder. If you are an agnostic, there is hope for you. If you are a hard-core atheist, I suggest that you read both of my titles, *Divinely Inspired* and the *Messiah Interviews*, and then temporarily abandon your rational intelligence for faith and belief in God. You will pick up your logical mind after you make the transition to believing and trusting in the Creator. Of course, God gave us His gift of free will, so it's your choice. Thank God, I have made mine.

What is spirituality? I'll tell you what it is not. It's not religion, although I am proud like Einstein to be Jewish and a member of the tribe. In biblical times God used his prophets to convey messages to the forerunners of the Jews, the ancient Israelites who were in a constant struggle to not fall off of God's spiritual ladder. As the prophet Isaiah in Isaiah 29:13 says, *"These people draw near with their mouth, they honor me with their lips but their heart is far from me."* You can spout all the piety you want about how good a person you are and how you pray to God multiple times a day. However, if your heart is not a pure flame of spirit that is honest, true, tolerant and fair, then you are not spiritual.

In my humble opinion, religion does not offer salvation. Spirituality does. Religion can be a part of our spirituality, but not the other way around. There is so much grief in the world because of religion. We cannot solve the global crisis of spirituality and bring peace to the world because everyone is operating with a different playbook, a different deck of cards. What we can do is take the road less travelled like Einstein did and allow the spirituality buried within each of us to emerge.

We need to ignite our *pintele yid* in Judaism or our *kernal* in Buddhism. God has purposely hidden this spark in our Divine soul energy located in our hearts so that we can have the joy of discovering it on our own. If you have ever felt that you are leading a meaningless life and that you are not important, like a bump on a log, you won't any longer if you ignite your flame. It will give you something that money can't buy, and you will have a lot more than wealthy people without spirituality. As the great Chabad Lubavitcher Rabbi, Rebbe Menachem Mendel Schneerson, wrote, *"God did not make one useless thing in this world."* Also, you add to the world by being spiritual. Then you can be a truly good person. The Rebbe again: *"All that add. They add to Him."*

Spirituality is therefore Divine soul energy inside of you that allows you to act with a humble heart and be a kindred spirit and beacon of hope for those within and outside your inner circle of family and friends. Spirituality is an appreciation and a gratefulness for life and the gifts that God gave us in the world we live in. When you have it, you don't have to flaunt it. It's inner peace where you are defenseless, non-struggling, and tensionless like a river flowing and bouncing off the rocks in the water or like the deer that frolic hither and there. It's a respect for every living creature and for nature. If you attain spirituality, then you become more precious than gold. Some who have a natural spirit like my wife, Marcia, are indeed fortunate. Others like myself have to constantly move toward spirituality, brushing aside the thorns blocking my path along the way.

When you have climbed high enough on the ladder, you realize that it's easier to climb right back up when you fall off the ladder. Belief in God helps you on your journey because if God recognizes your efforts at righteousness, He is right there with you assisting you after your slide backwards. You taste of His wine and not only His fragrance and you have a right to expect His aid if you call out to Him. However, you can attain these elevated heights even if you don't believe in Him. The consequence may be that you will smell His fragrance but not be offered His wine and true guidance and help. God is omniscient, all knowing, and sees through the Divine souls he gave us. Our souls act like a candle by which God examines the light of our inner selves.

When you are spiritual, you see people without spirituality differently and like the river pushing away from the rocks, if their arrows are aimed at you, they are deflected. You also unexpectedly find yourself being with and enjoying people who also have discovered the golden elixir of spirituality. The old saying is true, *"Birds of a feather flock together."* As the days, months, and years pass, you see yourself differently and you grow wiser. Job said it best: *"Days speak and a multitude of years teach wisdom."*

If you are a believer, you see God in the shadows and know that your destiny is in His hands. When things are good, you praise and honor God, and when they are bad and you are crushed and beaten, you still sing to Him.

How do I begin? A plethora of spiritualists out there have how-to guides, so I'll just give you the direction that I took for my climb up the spiritual ladder. I have been at it about twelve years ever since Divine miracles entered my life. I have been so fortunate to receive the Creator's splendor. I feel honored for God coming into my life. In the Torah, God tells Moses that He will offer His blessings to anyone He pleases, even to a sinner who does not deserve it. I am and was that sinner.

1. Your first challenge is to abandon your rational intelligence if you have no faith. This is probably the most difficult step since God is a complete mystery to you. Once you make this leap, your intelligence can stand alongside your faith. Famous scientists like Einstein and Newton did it. I

too am a scientist. However, because of Divine Providence (Divine miracles), I did not have to abandon my rational intelligence. I immediately became a believer. If you can't do this, then just resign yourself to being an agnostic or atheist and do the best you can at being a humanist. My own feeling about agnostics, despite what they say, is that they are atheists with subliminal fear of God. They will cry out to God at times of extreme crisis. The hard-core atheists are not able to do this because they cannot swallow their pride or are anti-God because it's the only way to rationalize away their hidden sins.

2. At the beginning and even when you are halfway up the spiritual ladder, you will have doubt. Make it easy on yourself. Don't initially attempt to resolve your doubt in a church or synagogue. Instead, start to talk quietly aloud to God on solitary short walks. The Bible gives ample evidence that it's OK to be respectful and at the same time challenge God. Do not expect answers in hearing His voice, like I was so fortunate to hear. However, if you do hear His voice, consider this a rare blessing. Stay the course even if you see no change in your life. Ultimately, I believe you will and when it happens, you need to recognize that it is a gift from your Creator. In the Bible, King Solomon says, *"God will return to you if you return to Him."* If God sees some future potential in you to becoming spiritual, even if you are not right now, He will answer you.

3. If you don't see any changes say after three months or less in what you are asking of God, then you need to begin to take steps to improve your moral character. You need to discover and ignite your soul's *pintele yid*. It is close to being extinguished. You can begin to act righteously. God

is at all times giving us moral tests and He is keeping a scorecard of our decisions. Those decisions will determine whether we are admitted into the Messianic Age at the End of Days, as I outlined in my spiritual novel, *Messiah Interviews*. We all know the difference between right and wrong. It's choosing right over wrong. Often we cannot see past our feelings. We can't help that because of our upbringing. Be conscious of the following improvements to get your climb up the spiritual ladder activated:

i. If no one is hurting you, you have no right to hurt that person.

ii. Absolutely no one has the right to physically, emotionally, or mentally abuse anyone else—your spouse, your children, and so forth.

iii. Look up the Ten Commandments (see my books) and try to follow them as best as you can.

iv. Be charitable. If you do not help the pauper when he cries out to you, God will not hear you when you cry out to Him. Actions speak louder than words, so carry out some action even if you don't totally mean it. Gradually, you will feel good about giving to someone. If you cannot afford giving money, then offer kindness through your words or a smile.

v. Try to be more God-centered instead of self-centered. Look to the old adage. Light a fire to warm a room instead of wearing a fur coat and warming only yourself. Do some kind deeds. Show mercy and be compassionate.

vi. Be honest and sincere in your heart. God sees your intent. He values truth in your heart more than anything else. Don't cheat or steal in business or in life. Be an honest person.

vii. All God wants is to be recognized that He is our One and only God. Honor and praise Him by thanking Him, and be grateful for the gifts he has given you even if it's your waking up each day.

viii. Treat your fellow man with respect unless you know that person to belong to the Devil and to be evil. You don't have to like a person. You just have to be courteous unless that person is abusing you. Then by all means say something that provides him or her with a message to back off. Obviously, this won't work on a global scale where man's treatment of man is abhorrent.

ix. Be grateful for what you have. We often look somewhere else while we have it right in our own backyard.

x. Pray in silence when so inclined. Continue to pray aloud. God hears you either way, but there is something cleansing about expressing aloud.

What are the benefits of spirituality? All of the above suggestions seem simplistic, and they are. They worked for me. Honestly, I feel that the rest that you do, such as going to a church or temple and praying, is the tinsel. Practicing the religion when you are spiritual is a good thing. What deters us is what Shakespeare wrote in Macbeth a long time ago, *"The attempt but not the deed confounds us."*

When I was about 80 percent up the spiritual ladder, I believed that there were no coincidences. Everything that happens to me happens for a reason, and God determines my non-moral outcomes when I act righteously with justice in my moral decisions. For example in the Bible, Jacob prepares for his meeting with his enraged brother Esau. Yet He knows that the outcome of his rendezvous is dependent on God. Jacob never took his righteousness for granted, and you won't either, when you feel that inner flame of spirituality. I know how hard it is get the messages of my books out in a sea of books and a bad economy. I also know that my success is dependent upon God. And I'm not measuring success by fame and fortune.

There is an old saying, *"God helps those who help themselves."* In life I do everything I can on my own and then when I am absolutely stuck, I go to God and ask for His help. When I became spiritual, I was glad whatever way a situation turned out because that's the way I now believe it's supposed to be. I never appreciated nature, but now I love to walk in the nature preserves near my house in Florida. Being spiritual means I want to make decisions that God would approve of. It also means that I have an unshakable faith whereby I not only know that God can cure me of an illness, but that He *will* cure me. The verb tense is important. I have converted uncertainty to certainty in my faith, trust, and love of God.

They say that God can save you from the forces of nature, but He does not interfere with man's actions against man. Yet I have done womb regression therapy and God saved me from an abortion attempt. It's funny that the higher your personal desire to ascend the spiritual ladder and the higher you climb, the more you stumble. Ironically, the more you stumble, the easier it is to get up again. God has the most demanding standards for you to gain a foothold on the highest rungs of the ladder. Only a few in our history have been able to reach such heights. The Patriarchs,

Abraham, Isaac and Jacob, Moses, King David, and Jesus found themselves at the top of the ladder. There were others in the non-monotheistic religions.

This simple path that I have outlined is not an easy one. You have to invest more than your deeds to become spiritual. You need time, energy, and endurance in a life that is most certainly challenged to its full capacity. Those who reach heavenward may sow in tears, but they will reap in joy climbing the spiritual ladder. There is a biblical quote that I love, "God offers His fragrance to all but His wine is reserved for those who are spiritual and wish to be guided by Him." With spirituality you gain self esteem and self worth so that you can accept yourself and not look to others for your fulfillment. Most important, you will reserve your spot to enter the future Messianic Age where you can live for all eternity in a purified new physical world.

11

❧

DIVINE SOUL, ANIMAL SOUL, & EVIL INCLINATION

ROM NEWTON'S STUDIES OF THE SOUL WORLD, WE KNOW THAT
THE DIVINE SOUL INTEGRATES INTO THE FOETAL BRAIN. From
my own energy measurements, the Divine soul is located
over the left hemisphere, inside the right hemisphere, and behind
the eyes. The latter would explain the movement of the eyes in
Marcia's picture by Marcia's soul in eye to eye recognition. A low
percentage of the energy of her soul has decided to locate in our
home in Boynton Beach. As Newton discusses in *Journey of Souls*
and *Destiny of Souls*, any percentage of the soul is the entire soul.
It's still the soul no matter how much energy is associated with it,
like the analogy we made of the entire ocean in one drop of water.

It is interesting that the energy moves continuously from the
left brain to the right brain to the left eye to the right eye and back
to the left hemisphere. The trajectory of this movement roughly
follows the shape of the bulbous pear description of the soul in
the spirit world. The velocity of the energy is variable and the
direction is always in a counter clockwise direction in the brain.
If I find the Devil trying to interfere with my measurements, the
energy moves in a clockwise direction. The Devil particularly
early on in my measurements made it very difficult for me.

Within the left brain we find the evil inclination, which is under the influence of the Devil. The evil inclination is that part of your brain that is always negating or trying to negate your better beliefs and your better actions. Often it will change your mind for the worse and make you do the wrong thing.

Both the Divine soul and the evil inclination exist, though anatomists cannot find the location of these entities in the human brain. It is important to remember that the Devil exists at God's pleasure even though the Devil believes he is totally independent. God has given the Devil a tremendous amount of latitude in being able to exert his evil actions and to buy the souls of individuals who will carry out Satan's plans. You can detect these evil souls through measurements of your Divine soul, but don't expect these evil souls to repent. They might not repent even if they found themselves standing in front of a firing squad. What about the animal soul? Here is what I wrote in my book, *Messiah Interviews*.

The Garden of Eden was a time when no badness or evil could be found in God's Creation world. With the sin of Adam and Eve, evil pierced the human soul. Before this original sin, there was only one soul, created by God's breath into Adam's nostrils: the Divine soul.

By the Devil serpent's enticement of Eve with the fruit of the Tree of Knowledge of Good and Evil, another soul entered humans. Orthodox Judaism refers to this soul as the animal soul. Adam and Eve were commanded by God to leave the Garden of Eden; when they did so, both the Divine soul and the animal soul existed within them, as it does in all of humankind ever since this original sin.

As the very first ChaBaD Lubavitch Rebbe, Schneur Zalman, wisely suggested, "All history, since the beginning of Creation, has been about discerning evil from good and separating the two." Further, God gave man free will to choose good over evil by calling upon the goodness in his Divine soul over the badness in his animal soul. To solidify this concept in Adam and Eve's twin sons, an evil Cain utilized his animal soul to murder his brother Abel because God praised Abel over Cain for their offerings.

You might ask why God would create a world with a mixture of the two, good and evil. Why not just stay with the goodness of the Garden of Eden? Before we can answer this question, we need to clarify the nature of the animal soul. Is the animal soul all bad? No, the animal soul is a mixture of good and bad, in contrast to the Divine soul, which can be tapped only for its goodness. Like strumming King David's harp, the more you awaken your Divine soul, the more God's Light shines upon it. Goodness finds goodness, and the Light filters down to the good part of your animal soul.

The good part of the animal soul can be smothered by the bad part, and a person can become evil if he already wasn't evil to begin with. If all good is eliminated, we probably are dealing with someone who is demonic who commits mass murder of innocent people or has sold his soul to the Devil. Yet very bad people, in rare instances, have had epiphanies in which they alter their course in life, repent their sins, and become virtuous. This statement does not apply to the Devil's own.

Examples of the good part of the animal soul acting in the absence of the Divine soul could be taking your children to soccer games, buying your wife flowers on Valentine's Day, contributing to society by working, following your religious beliefs, being joyful on the Sabbath, and so forth. In short, anything that does not involve a moral choice.

Once a moral choice to choose between right and wrong comes into play, the Divine soul in your brain either kicks in to pass God's Light into the good part of the animal soul to choose good, or the bad part of the animal soul is activated by the evil inclination in the left brain to choose bad, or worse, evil. Charitable gifts to help the poor and the needy or choosing to do a kind deed for another person without expectations for your own benefit involve a moral choice to do so.

Orthodox Judaism speaks about an evil inclination, or yetzer hara, and a good inclination, or yetzer tov. God holds us accountable over the course of our lifetimes for the moral choices we make with our gift of free will. He will sit at the End of Days and judge each one of us, who will be resurrected to live eternally in the Messianic Age. He will assess us for our moral strengths and weaknesses.

There is no doubt that the Divine soul is what allows God to be the keeper of the scorecard. Our Divine soul is also part of Him. How God tallies up the scorecard for righteous or sinful acts is anybody's guess, because now you are within the realm of the unknown, God's essence. As God clearly tells us in the Hebrew Bible through the prophet Isaiah in Isaiah 55:9, "As the heavens are higher than earth, so My ways are higher than your ways and My thoughts are higher than your thoughts."

How can atheists or even agnostics receive Divine Light if they don't believe in a Divine soul? We just can't leave out those humanists who are walking with God. They too deserve entrance into the Messianic Age. You might counter with, if that is the case, then according to your reasoning (mine and Marcia's), there seems to be no advantage to living a spiritual life.

We believe that humanists only receive God's fragrance. To receive God's wine, you have to believe in Him, although your actions must be equally righteous, quality-wise to the deeds of the humanist. Agnosticism, without righteousness, doesn't qualify for admission into the second Garden of Eden at the End of Days. Neither does righteousness or religious piety without sincerity of the heart.

"Where is the location of the animal soul?" It is on the right side of the chest directly opposite the heart. It too cannot be found in the body since it does not occupy space, having been placed inside us by God. Checking our soul energies, Marcia and I have determined that our animal soul can be represented by one hundred receptors. We all start out with 77.5 of these receptors lit up as Light and 22.5 receptors darkened as Dark. Imagine that these receptor are little hollow balls attached and integrated into to a larger sphere, the animal soul. At the end of your life, if you still have the 77.5 Light balls or more, you will be admitted by God and the Messiah into the bliss of the Messianic Age. It's like getting a B average or better in order to gain admission.

The Light and Dark of the balls can fluctuate during the course of your life and can individually be filled as a mixture of

Light and Dark depending upon your moral choices. If you falter and steal monies from your company and go to prison, many of your Light balls may darken. If you come out of prison, you still have the rest of your life to atone for your sins and do good deeds. However, if you commit murder, all of your Light balls will likely get wiped out and you may never have a chance to recover. Good deeds act in the opposite way as the Dark balls lighten up wholly or partially. Going to a food shelter and helping out will help you increase your Light balls. When you are confronted by an alternative moral choice, like stealing or intentionally hurting someone physically or emotionally, choosing not to steal or hurt will turn on Light balls. Since the Light balls are really the Light energy of the Divine soul, God knows your status at any point in time. Marcia was always way above the 77.5 mark, but not me. It's only been the last twelve years where I feel comfortable with who I am. No one on earth has his one-hundred receptors fully lit up since God made human beings imperfect.

When I was a professor in the oral biology and pathology department in the Dental School at Stony Brook University, I taught the students how we secrete saliva. During the physiological processes, the salivary glands have receptors on them that get activated by neurotransmitters travelling along our involuntary or autonomic nervous system. There are for the most part two kinds of receptors, those activated by our parasympathetic nervous system and those activated by our sympathetic nervous system. In what we are describing here, there are also two kinds of receptors on our animal soul, good Light receptors and bad Dark receptors. Instead of the sympathetic and parasympathetic nervous systems originating out of the human brain and spinal cord, we are suggesting that special invisible energy pathways come out of the evil inclination and the Divine soul. When a bad or evil deed is chosen, the energy comes out of the evil inclination, and the Dark receptor balls get activated, allowing the animal soul to perform the bad deed or bad choice. When a good deed is chosen, the Divine soul energy activates the Light balls to perform the good deed or make the good choice.

This seems all too simple. However, remember it is the energy that is released by either the Divine soul or the evil inclination that is activating good and bad receptors. Remember also that it is God who gave us both the Divine soul and the evil inclination, but it is the Devil whom God has chosen to influence our evil inclination. That voice in your brain is really the Devil and never take him for granted. The Devil is extremely powerful and is the epitome of evil. In some individuals the Devil owns their souls virtually becoming the evil inclination itself and thus he totally controls them. They have no free will of their own and are at all times under the Devil's nose and control. It is difficult to conclude this because the Devil does not activate the bad part of the Animal soul in these evil people and they therefore appear as nice and good as apple pie and you would never know their evil if you met them or even knew them well. They would be under your very nose and you wouldn't have a whiff of their evil intentions to harm you.

Therefore we all start out with the same energy potency, but we all have distinct souls that keep changing with past lives. Hopefully, we can refine our souls in each life body that we incarnate into. God, as we mentioned, also may select certain individuals for particular tasks on Earth, such as Ghandi, Martin Luther King Jr., Winston Churchill, and Harry Truman. For the most part, though, greatness is not innate to the individual. These men used their God-given potential and rose to greatness. They were there at crucial times when there was a tipping point or breaking point of humanity. However, God will always ensure the survival of humanity because his plan is to go from an imperfect world to a more perfect world of the Messianic Age. The hidden lesson of the first Garden of Eden was that Adam and Eve were held accountable by God and were booted out because of their choices. We too are accountable and God is watching us at every step.

The Hebrew prophets in the Bible talk about God giving those who qualify for the Messianic Age a new heart and a new spirit. What does this mean? Well, for one thing, the sages have spoken about a future Redemption when the evil inclination would be subdued although free will would still exist. To Marcia and me, this would represent a new jump in evolution, creating a different kind of human being. We call this new species "spiritual human beings." Free will would still allow a person to screw up, but if he did, he would lose his ticket to eternal life.

We said that the heart contains the *pintele yid* soul energy, which is normally hidden and needs to be discovered by each of us. We have to spark our *pintele yid* to begin our spiritual journey. In Messianic times, this holy place in our hearts will no longer be hidden. This is what it means when God will give us a new heart. The new spirit can only mean a new soul. Marcia just told me what this means. The new soul will allow us to be in close contact with God on a more personal basis. We define all the progress of climbing the spiritual ladder as *becoming closer to God*. The Bible tells us that at the time of the world's redemption, we shall know God as the waters of the sea. We may even get to see Him transform Himself into human form.

"How many energy particles are there in a Divine soul? Are there one hundred? Are there one trillion?" When a new soul incarnates into a new body, it gets a new piece of God and also carries all the tiny bits of the souls from past lives. We all seem to get about the same number of energy particles, though depending upon where we are on the spiritual ladder, God can add energy particles. We can also give away some of our Light energy particles to heal another person, and good deeds are rewarded with more Light receptors, lighting up the dark receptors on the animal soul and more Light energy particles. "And what holds true for the Dark energy particles of the evil inclination? What good are energy particles anyway?" Wow! What a barrage of questions that Marcia needs to supply answers for. Let's take them one at a time.

Marcia answered the first question, and I confirmed it was her and not the Devil and also the number she assigned. Seventy-five million energy particles make up an infant's soul at birth. These energy particles get distributed throughout the body at birth and may increase as the infant grows, ages, and the individual acts righteously and kindly. At birth, these energy particles are compacted at the top of the left hemisphere, inside the right hemisphere and between the eyes. I had to determine the next answer, the diameter of one Light energy particle. I came up with one millimicron, which works out for the soul of seventy-five million particles to be a total of seven-and-a-half centimeters or three inches. "How many Dark energy particles are in the evil inclination?" There are a smaller number of dark energy particles, about one million. "Are the Dark and Light energy particles equally potent?" Yes!

"Finally, what do these good energy particles do for you?" First, you need them to be alive. The body needs the soul to live. Second, you need this Light energy to activate the Light receptors on the animal soul. Third, the intensity and the number of the Light energy receptors of your Animal soul increases as your spirituality increases, allowing you to subdue or eliminate your evil inclination. God is most interested in you overcoming your evil inclination, and He has given you the tools to do so — free will. You might ask, "What's the point of subduing your evil inclination if you are going to die anyway?" The brass ring is that you are admitted to the Messianic Age to live in a second Garden of Eden for all eternity.

We can determine the number of energy particles in the Divine soul, but can we calculate the number of energy particles in God? How does God fit into scientific theories of our universe.

12

❧

GOD & THE BIG BANG
CREATION & EVOLUTION

I N READING ABOUT THE ORIGINS OF OUR UNIVERSE AND OUR
EARTH, A CONCENSUS OF SCIENTIFIC OPINION HAS BEEN BUILT UP
TO DESCRIBE THE EVENTS THAT OCCURRED.
Astronomers believe that billions of years ago, all matter was
concentrated in a single mass. We have suggested by our energy
measurements that this highly compressed mass was God. We
further have suggest that God, Who was Light energy particles,
contained the darkness and all the constituents necessary for the
early universe. We call this a reverse black hole since usually
darkness consumes light, where here we are proposing that Light
trapped darkness. This is the Light and darkness described on
the First Day of Creation, since the sun and the luminaries did
not appear until the Fourth Day. God's Supernal Light was not
released until in a moment of God's choosing, the Light emerged
from its resting state and there was a "big bang," an explosion
where scientists hypothesize that the concentrated mass blew
apart. Scientists have estimated that the universe was at least a
thousand times smaller just after the big bang than it is today,
as space has expanded. The explosion caused God also to expand
His Light and even when the universe was just seconds to a

minute old, God released hydrogen and helium gases into the atmosphere.

"In time scientists believe that a disk-shaped cloud of dust formed the sun and peripheral matter formed the planets. At the same time or earlier, primordial gases condensed to form the first stars." According to biblical creation, the sun and the luminaries in the sky did not appear until the Fourth Day, but millions to billions of years had passed from the First Day until the Fourth Day. God gave us the term *day* when He described Creation. Since time has no meaning for the Creator, His purpose was to introduce structure into our world. The seven days of Creation became our week, and God's resting on the seventh day gave us the concept of our weekend. While God is not governed by time or space, we humans are. Even the sun is subject to time constraints, rising every morning and setting every evening.

"As time marched on, heat produced by compacted radiation and impacting meteorites then melted our planet earth. As our planet cooled, the Earth's layers formed. During this cooling, water vapor condensed and torrential rains filled up its basins, forming the seas." The first signs of life on Earth 3.8 billion years ago were microorganisms. But no one has figured out where these bacteria came from. It is our belief that angels flying spaceships brought these bacteria to Earth so that God could jumpstart evolution. Nobel prize winner Francis Crick many years ago proposed that aliens brought the microorganisms.

As you see, we have inserted God into the narrative. If the Divine soul at birth is seventy-five million energy particles, "What was the quantity of energy in God before and after the big bang?" Before the big bang, God was one million quadrillion particles, which would represent a diameter in space of 40 trillion inches or 3.3 trillion feet or approximately 625 million miles. If the universe expanded approximately one-thousand fold from right after the initial expansion of the big bang until now, God's energy would also expand, and today it would represent at the low level one billion quadrillion particles contained in or scattered in 625 billion miles of space. This if true would be a huge number of miles and would be a huge number in line with scientific views that trillions of miles of space are thought to be a crude estimate of the distance spanning the universe. God represents a large part of space and He has the potential to spread His energy particles throughout space in a God-Field so that He essentially is space itself. Therefore there can be so such thing as nothingness in space, although God is invisible to us. Who would believe such a preposterous notion? Albert Einstein would at least be open to the possibilities. Furthermore when we practice Tai Chi or Yoga, we can absorb God's energy not only from the atmosphere but also from the earth because God's energy is everywhere and is universal.

In the Bible, God tells Moses that no one can see His face lest he dies. If God represents one billion quadrillion energy particles as we are suggesting, then our gazing at God's full brilliance would be far too much for us to bear. In Messianic times, God would either have to dramatically reduce His Light energy or we would have to look at Him not just with our eyes but with our "soul eyes" if we are to visualize Him. Furthermore if God occupies at least 625 billion miles of space, we might ask where is Heaven? We know that the sun is 93 million miles away. According to Marcia, Heaven is a thousand times further, 93 billion miles from earth. This is not too far for the energy of our souls to travel at 186 billion miles per second but it is for our fastest spaceships where speeds currently approach only 40 thousand miles plus per hour. Marcia and I have confirmed that our thought-energy transmissions indeed reach each other in one-half a second.

Using the world population of seven billion, with each soul being seventy-five million energy particles, God would need to use five hundred and twenty-five million quadrillion energy particles, about half, of his total one billion quadrillion plus energy particles. Because the universe continues to expand, God's energy and mass are also expanding so that there will never be a danger of running out of sufficient energy to accommodate all the souls on the planet. As you can see, our soul energy is just a drop in the bucket in comparison to God's energy. If you think of the United States National Debt at 15 trillion plus dollars, this is how much more energy God has than any human being, 15 trillion times as much.

One of Newton's patients had talked about the fact that the cycle of thousands of years of incarnations into new bodies would end at some point. The Jewish Talmud written fifteen hundred years ago states that our current Earth will end six thousand years after the Creation of Adam and Eve. Since no one knows the exact date of Creation, and since there has been a loss of 164 years in the Hebrew calendar, we may be close to the six thousand year mark. God therefore looks like He has timed the ending of our world and the beginning of the Messianic Age World to Come perfectly.

From our discussion, it seems like Marcia and I are flippantly commenting on evolution that God jump-started 3.8 billion years ago with the first bacteria and the Creation of Adam and Eve in the Garden of Eden six thousand years ago. We are not. One of us, Jerry, does have some credentials. I was, am, and will always be a scientist, and therefore I enthusiastically acknowledge that man existed before the Creation of Adam and Eve almost six thousand years ago. I also have an unshakable belief in God, so I am a man of faith as well as a man of biological reason and logic.

The one important opinion that we are missing, and that is not an opinion at all, would be God's, if He chose to take the public stage as He once did back in biblical times. Back then God was conversing with Abraham, Isaac, and Jacob and with Moses, Kings David and Solomon, and the Hebrew prophets. Certainly God could shed His Light on explaining what really went on including giving us an exact quantitative value for His energy..

God's earliest appearance as far as we know was in the Garden of Eden with Adam and Eve and the Serpent, Satan. However, God began our universe approximately 14 billion years ago, our Earth 4.5 billion years ago, and as we just discussed, He jumpstarted evolution 3.8 billion years ago, when we see the first fossilized forms of microorganisms.

Einstein thought that you cannot solve a problem at the level the problem was created. There is no doubt from watching Creationists, with or without intelligent design supporters, battle with proponents of evolution that the controversy over Creation and evolution seems to be created at the human level. Therefore, according to Einstein, we need to solve it at another level. The only level I see that can solve the problem short of God Himself is the Divine level.

We are proposing that God created the problem in the first place. In our humble view, God is the master scientist responsible for both evolution and Creation. He is also an all-powerful omnipotent God Who is all-knowing or omniscient. The immense quantity of his energy particles spoken about allows God these attributes. These energy particles represent an organizing intelligence that we cannot begin to comprehend. The soul energy must be a mini brain, while God is a super nova maxi brain. This is the inescapable conclusion. Otherwise you cannot explain God's omnipotence-omniscience and on a much smaller scale, the telepathic communication between Marcia and myself.

To reiterate the Bible quote, "*As the Heavens are higher than the Earth, My ways are higher than your ways and My thoughts are higher than your thoughts.*" We understand very little of the essence of God and to even come close to understanding something of His plan for humankind, we have to abandon our human brains and try to think like God. Of course, this is an impossible task as we are limited in all that we are and do, while God is limitless.

The "Out of Africa" evolutionary hypothesis is that human life as we know it since *Homo sapiens sapiens* was first seen on

Earth, perhaps fifty thousand to one hundred thousand years ago, is God's scientific experiment. The objective of the experiment was to create an imperfect world where humankind could use the gifts of free will, language, rational and random thought, intelligence, wisdom, creativity, imagination, fantasy, and doubt to lead an interesting life, hopefully with purpose. The Garden of Eden represents the affirmation of good and evil, although even before Adam and Eve, when men were living as marauding bands of cavemen on the savanna, they were at each other's throats. As God puts it to Noah in the Bible after the Great Flood, *"The imagery in man's heart is evil from his youth."*

To know perfection, you need to experience imperfection, just as feeling hate really prepares you to love or feeling pain allows you the ecstasy of pleasure. The same is true for happiness and sadness. The perfection that God has planned for us is a second blissful Garden of Eden, the Messianic Age at the End of Days. God's scientific experiment was therefore to create an imperfect but dynamically challenging world so that in the End Times, when there would be a Christian or Jewish Messiah, we would be part of a more perfect world.

Arguments about Creation versus evolution are actually a good thing and fall into the goal of God's plan of creating an interesting world. However, we need to just accept one another's personal views at this stage with the knowledge that we'll come together for explanations in the Messianic Age. In the Hebrew Bible or Old Testament, we can reach the Messianic Age through a peaceful transition, an apocalyptic ending described in vivid detail in the Old and New Testaments, or by the arrival of a Prophet. Disputes about evolution and Creation fall into the peaceful category, providing we have a disputation or friendly disagreement. When we condemn one another with expletives and disrespect one another, God scores that in favor of a violent ending for the world. If you realize the overall goal of the Divine, you can even bring a sense of humor to the table. We need to go beyond the dogmatic and aspire to tolerance of one another. We don't have to like one another. We can, however, despise Satan and his evil followers.

I wrote about my thoughts in *Divinely Inspired*. Here is an excerpt from the epilogue: *"Although the fossil record is scanty, there is no doubt that we Homo sapiens sapiens existed long before Adam and Eve. Over the last twenty years, there have been discoveries of human skulls such as Kenewick Man, who lived in the United States about ninety-four hundred years ago, almost four thousand years prior to Adam. And we know that our ancestor, Cro-Magnon Man, who is us, was in Europe 40000 B.C.E. and died in 10000 B.C.E., perhaps due to the coming of an ice age period. From the cave art found in Southern France in 18000 B.C.E. and the engravings even before, it seems clear that Cro-Magnons had language and a God-given soul. Cro-Magnons were followed by Neolithics, who essentially were hunter-gatherers of wild foods and who were acted upon rather than controlled their environment.*

Given the above, why did God create Adam? We need to continue the story a bit further and turn to a description offered by Isaac Asimov before I can offer you my opinion. By 8000 B.C.E. in Mesopotamia, which was in present geography 70 percent in Iran and 30 percent in Iraq, humans had learned to plant seeds and to gather the grain that grew from the seeds. This was the beginning of the Neolithic Age or Stone Age. Better and more polished weapons were produced, pottery was developed, and farming and the herding of animals became the livelihood of the day. The result was a population explosion, because food became secure and could be stored.

The hunter-gatherers disappeared and gave way to communities of villages, and civilization began to develop. Agriculture from the Iranian highlands spread to two areas in particular, the Tigris and Euphrates locale just to the south and to the Valley of the Nile in Egypt. All of this was in full force around 4000 B.C.E., as Neolithic villages flourished. As methods for farming continued to improve with increases in productivity, new areas were colonized and high population densities in towns and cities sprang up. The first civilizations in the world were very much present by 3500 to 3000 B.C.E. in modern Iraq and in Egypt and were followed by new urban centers in India, China, and later America.

Simply stated, God announced himself with Creation of Adam and Eve for the first time to the human race in around 3760 B.C.E. Why now? My guess would be this was an opportune time with the population increase taking place in developing and congested cities. No one can presume to second-guess the Almighty. However, it makes sense to me that this was the time for God to appear, since humanity would now be in close proximity to each other and we all know what history has shown us about humanity's interactions with each other. Even when humans were living as marauding bands spread out on the savanna, it was their natural violent instinct that led them to clash with each other. Now with urban crowding brought on by the advent of civilization, a public announcement of a Higher Being was in order.

Man's inhumanity to man was in evidence in the first ten generations of Creation Humans. God caused a great Flood to consume everyone except for Noah and his family around 2100 B.C.E. in a localized part of the world. In my opinion, it couldn't have been the entire world, since there were pockets of mankind developing via Evolution in other parts of the world at that time. Evolution was ongoing because God had initiated it. Not all of mankind came from Adam and Eve, but it is my view that at least the Jews were biblical descendants, as were the Christians and Muslims. Therefore, I believe God made His first public appearance at the time of Adam in order to show humanity a better way, but humanity didn't get it back then, nor have we gotten the message throughout history.

The message advocated by God through Moses, His prophets, and Jesus was that humans should live in harmony with each other and be as one with nature and the environment. God made quite a few more important public appearances during biblical times, but I believe that it was at the Revelation at Sinai with the giving of the Ten Commandments that God gave it His best shot to get His message across to mankind. Mankind at the time consisted of idol worshippers and Israelites. The Israelites got the message. They just didn't follow through or sustain goodness. The Israelites lapsed in and out of being idol worshippers and conducted themselves immorally against everything the Ten Commandments stood for."

With the quadrillions of energy particles that God possesses, He was able to speak in a loud voice to the Israelites at Mount Sinai when He delivered the Ten Commandments. At that time man's evil just could not be contained, and the Ten Commandments represented God's best shot to level the playing field between good and evil. Although the Israelites lost their way and did not follow the Ten Commandments, they, like we, cannot abrogate what God has commanded us to follow. Continually breaking the Ten Commandments will assure one thing: You will not be admitted into the Messianic Age.

God not only gave us the sperm and the egg to create a body. He also provided us with a life-activating Divine soul possessing intelligence independent of our human brains. Just how the soul acts as a vitality factor is not understood at this time. It's possible that one function of soul energy is is to take oxygen into the lungs of the newborn during the baby's very first breath of life. Another function might be that soul energy somehow is an active transport mechanism for oxygen into our cells.

Is God the Creator of all things? Science might argue that God plays no role in our universe but with the help of our friend Thomas Spradley, Marcia and I proposed in an article entitled "The Higgs Field, Higgs Boson Particle, and GOD: Are they one and the same?" that we published on the web in the Weissman Report. It is possible to blend science with spirituality. Here is the article we wrote.

Summary:

Every once in a while, an opportunity arises which permits us to blend science with spirituality. Such a momentous event was announced on July

4, 2012 with the discovery of the Higgs field and a particle thought to be the predicted Higgs-like boson particle originally proposed by Peter Higgs and five other scientists in 1964. Physicist Leon Lederman in his 1993 book, The God Particle, deemed the Higgs particle with its Supreme Being name because its elusive particle field is an unseen field that permeates all of space. Furthermore it is central to our understanding of the structure of matter, specifically how elementary particles are given mass. The term "God particle" was considered an exaggeration or hyperbole by scientists, but is it? In 2003 in his book, Divinely Inspired: Spiritual Awakening of a Soul, Jerry Pollock proposed that God was the master scientist responsible for both Creation and Evolution. With the publication of Putting God Into Einstein's Equations: Energy of the Soul in April of 2012, Jerry Pollock from our earthly world and wife Marcia Pollock from the spirit world have collaborated to spiritually calculate the God Field. We suggest the possibility that the "Higgs field" and the "God Field" could be one and the same.

The Science:

We briefly describe the science behind the hypothesis of the Higgs three-dimensional field that has a characteristic action when elementary particles like an electron or quark collide with its invisible shield in a quantum excitation. The Higgs field gets activated and particles including the Higgs boson are created and then very rapidly decay. It is through the analysis of the decay that physicists can deduce that the emitted particles were probably created from the Higgs boson and not some other source. To form the boson and other like-molecules, a massive particle accelerator like the Large Hadron Collider (LHC) at the CERN Laboratory in Geneva, Switzerland is needed to create collisions with sufficient energy.

Particle physicists examine elementary particles and forces which are believed to create our world. The behavior of these particles and their interaction have been explained currently by hypothesizing a Standard Model. However, the mathematics predicts that the elementary particles would not have mass, yet they do. How is this possible? Enter the Higgs mechanism and the need for a particle, the Higgs boson, which could explain how mass could be acquired by elementary particles. The particle generated at the CERN laboratory had a mass equivalent to 133 times that of a proton which is consistent with the massive Higgs boson. As the elementary particles migrate through the quantum excited Higgs field during fine-tuned symmetry breaking, they acquire mass. At the same time, the Higgs boson mediates the interaction of the elementary particles with the Higgs field. The immediate result of this mass acquisition is that the elementary particles would have difficulty moving through the Higgs field and would never attain the speeds necessary to move through the universe at warp speed.

GOD:

We suggested from our collaborative studies that there is a God Field and we now propose that it may be the Higgs field. The Standard Model suggests that the Higgs field is unseen. In Putting God Into Einstein's Equations, we proposed that at the time of the big bang, only God existed as unique Light Energy Particles. These resting Light Energy Particles acted like a reverse black hole and contained the darkness and all the gaseous ingredients necessary to kick start our early universe 14.5 billion years ago. At a time of God's choosing He caused the big bang, and today God's Light Energy represents one billion quadrillion of these special particles.

Spiritual energy measurements, confirmed by Marcia's powerful bodiless soul, suggest that God's energy particles occupy 625 billion miles of space. If true this would be a huge number in line with scientific views that trillions of miles of space are thought to be a crude estimate of the distance spanning the universe. God therefore represents a large part of space and He has the potential to spread His energy particles throughout space, so that He essentially is space itself. The God Field like the Higgs field has never been detected.

Divine Soul:

Divine soul energy can be measured within one's body by a technique described by Yuen. It helps to have gone through Tai Chi and Qigong to be able to feel this energy. If one looks at the four of spades and asks one's soul the question, "Is this the four of spades?" one will get a Yes answer and feel the energy circulating in a counterclockwise direction around your heart. If one while looking at the four of spades asks, "Is this the six of spades?" one will get a No answer and the energy will be felt below your navel. If one then looks at the four of spades and thinks that it is the six of spades and then asks, "Is the four of spades the six of spades?" one will get a Yes answer and the energy will once again be felt around your heart. The inescapable conclusion is that energy follows thought because once you thought you were looking at the six of spades, it became the six of spades energy-wise, even though you were looking at the four of spades.

Thoughts originate in the soul and not the human brain and Marcia and I communicate directly by thought energy telepathy at one million times the speed of light. Marcia is in Heaven 93 billion miles away and the thoughts of our souls reach each other in one-half a second in the same way that a legitimate psychic serves as a medium between love ones who are alive and dead. Souls in the spirit world are bodiless.

While elementary particles may slow down in the God Field or Higgs field, soul telepathy via thought-energy does not and warp speeds can be achieved. It is possible that God's special electromagnetic energy travels through space time through wormhole shortcuts. Wormholes were originally conceived in Einstein's equations of General Relativity as Einstein-Rosen bridges, and traversable wormholes have since been proposed as better time-space models for traveling through the universe.

Hopefully scientists will realize that when elementary particles communicate at great distances in quantum physics, they are communicating as God's energy. Are God's powers limitless?

13

⸺✦⸺

THE LIMITLESS GOD

USSELL STANNARD'S *RELATIVITY: A VERY SHORT IINTRODUCTION* PROVIDES US WITH INSIGHT INTO THE BRILLIANT MIND OF ALBERT EINSTEIN. Einstein's special theory of relativity basically stated that if you move at high speeds, time slows down, spaces squashes up, and you get heavier. Indeed you could be squashed flat without feeling a thing, and you would live forever if you flew at fast enough speeds. His general theory of relativity dealt with accelerated motion and how curved space-time changes our understanding of the universe.

Would any normal person believe that a ray of light could be bent? From Einstein we can see how God and the soul could live forever because He and the soul travel at unbelievable speeds. Only God can go back in time, if you exclude the movies like *Back to the Future*. For God, time is meaningless, as the past, present, and future collide as one instant immeasurable by time because God Is, as Marcia always told me.

In a sense we go back in time when we travel to our past lives. However, we do this through the energy of our living souls connecting with the energy of our past souls, both souls of which are gifts from God. From Einstein we know that energy possesses mass. In the case of our Divine souls, this Mass possesses intelligence and recognition so that we can communicate with the souls of our past lives. Marcia's soul is very wise, and I communicate with her in an intelligent fashion as she has a thinking soul brain.

Given that we think we have only our human brain to think with, no pun intended, Einstein's theory goes against all common sense. Einstein had an interesting take on the logical mind: *"Common sense consists of those layers of prejudice laid down in the mind before the age of eighteen."* If there is difficulty for the average mind to comprehend Einstein's physics, how can it ever be possible to understand God and the essence of Who He really is?

How could an invisible God have spoken out to six hundred thousand Israelites at Mount Sinai when He delivered the Ten Commandments? How could God have parted the Sea of Reeds or brought the Plagues on the Egyptians so that Moses could free the Israelite slaves during the Exodus from Egypt? We know with our human mind that a day consists of light followed by darkness and darkness followed by light. Yet the Plague of Darkness lasted more than one day in an equatorial part of the world where this can't be possible. Day must follow night and night must follow day.

I know what you're thinking. All of this is just crazy talk from Jerry, who has suffered psychotic delusions and hallucinations and heard "voices" because of the bipolar disorder he suffered in the nineties. Or Jerry is truly nuts because he is now collaborating with his dead wife Marcia's soul to write this book. And though she is sane, Marcia is perhaps as nuts as Jerry, both of them talking about soul energies and Divine energy of God when many people don't even believe that God exists.

The miracles created by God just didn't happen in biblical days when Joshua through the powers of God made the sun stand still or the prophet Elijah revived the dead. They are happening right now, even though God is no longer on the public stage as He was back in biblical days. God has been hither and there;

people have come forth and talked about their miracles, such as dying on the operating table but then one hour later becoming alive again. Others have had near-death experiences and lived to tell about it.

Scientists cannot adequately explain paranormal phenomena. Although a lot of whackos are out there, legitimate psychics can contact the spirit world and ghosts are flying around in haunted houses. Telepathy is possible and certain individuals can read our minds. UFOs do exist; otherwise why would the government be trying to hide their existence? What gives people these powers? Atheists will tell you that anything outside of our own realm of reality is nonsense. Don't listen to them. God created a dynamic interesting world and gave us all of these things by providing us each with unique souls that possess an intelligent energy. God has unified our universe as one large whole where everything is in a dynamic state of energy flux.

Our souls give us the special sense of time that Stannard refers to as the "Now." Just like the soul creates our thoughts that happen so rapidly, we instantaneously proceed "from one Now to the next Now." We agree with Stannard that our consciousness is acting as a searchlight scanning progressively the time axis to label that special moment the "Now." Marcia and I attribute this ability to our soul and soul consciousness, not to our human consciousness.

Marcia and I make our mad statements with *chutzpah,* which means we have the gall to speak to you with a straight face while we in our case are telling you the truth. If you can lay aside your human brain while reading our book, you will conclude that in God we are dealing with "Someone" entirely different than anyone on the planet. He is an incredible limitless omnipotent, omniscient, omnipresent Creator who was kind enough to give us our world and life itself. For Marcia and me, we are truly grateful for all the miracles that He has bestowed upon us.

Marcia, our daughter Erin, and me each have heard God's voice and both Marcia and I have had conversations with the Creator. Mine have been more recent during the last few months during hypnotic regression, but there was one conversation that Marcia had a long time ago, which she told me about only now as a soul in the spirit world and not in our life together on Earth.

Marcia's tumor in the right lobe of the liver was the size of a watermelon. After her death I started to think about how many years she was carrying the cancer, given the fact that cancer cells can multiply and double very slowly compared to certain bacteria, whose doubling time might be every twenty minutes. Some cancer cells double only every three months, and the tumor in the right lobe was in the order of a half a billion cancer cells.

I began to think that perhaps Marcia had her cancer way back in the nineties. The question rolling around in my mind was, did she know about it and did not say anything? When I recently challenged her on this, her soul admitted to me that this was indeed the case. I suspected that it was 1995 when God came to her, because something odd happened back then. When I asked her about the date, she confirmed my suspicions.

I remember the scene vividly because throughout my life when something strikes me, I do a double take. This is followed by an imprint into my brain that I internalize and store until I need to recall it, even if the scene dates back as far as the womb. I have excellent past memory. I had just been released from the hospital after a second suicide attempt. Marcia and I were in our living room in New York, and she was telling me something I definitely didn't want to hear and had no defense for.

"Jerry, I think we need to split up." My response in my severe clinical depression was "Marcia, I'm in no position to help myself. You do what you need to do." Marcia left me immobilized and dumbfounded and disappeared into the bedroom. It was a short time later that she returned and said, "No, Jerry! We shall get through this together," or words to that effect." In *The Wing of the Butterfly*, Marcia's memoir, which was just completed for family and friends, I wrote that Marcia had wanted to separate because of all the trauma she went through in her life. Just when she thought she had found stability with me, her world once again came apart. However, as I was to learn now, this was not the whole story.

At the time I found what transpired between Marcia telling me that she was going to leave me and then saying she wasn't as one of those striking moments that my brain would internalize and that I would later remember sixteen years later. Speaking as a soul now in the present, she told me that she saw no hope for my coming out of my agitated depression.

When she left me and went off by herself into the bedroom, it was then that God came to her and spoke with her. Marcia asked God for strength and told God that if He would save me, she would die for me. God said that it was destined that she would have cancer, but because of her willingness to sacrifice for me, she would be given extra years of life. He said that her death was necessary to bring about the Messianic Age. In some ways her death was like the matriarch Rachel, who died in her thirties and would return to cry from her grave on the road to Bethlehem for the Jews when they were exiled from the Jerusalem vicinity to Babylon by King Nebuchadnezzar.

God did not speak further about the nature of the cancer, and Marcia had no idea that it would be liver cancer or when it would happen. Of course, as she grew more tired and preferred to lie on the couch and drink her tea, she sensed that the time of her death was coming, although she never let on to anyone. That was Marcia's nature. Given that the cancer had metastasized to the bone, the doctors could not believe how well Marcia looked as the time drew near for her passing. She was very thin, but in my eyes she was still beautiful.

Marcia made the decision back in 1995 to self-sacrifice for me after saving my life in my first suicide attempt three months earlier. I wish it was me that died and not her, but God had His own plans for both of us. To compensate, God has given me the gift of her spirit. Yet you always want more and I ask God through Marcia's soul pretty much every day to resurrect her so that I can hold her in my arms once more. Marcia's answer to me is always, "Soon" and she can say no more because only God knows when the Resurrection as part of the Final Redemption will take place.

God has such miraculous powers and such magnificence that it is easy to understand how He designed our universe. With the expansion of the universe, God's mass of energy particles too expanded and therefore His energy and speed increased. With continued expansion God grows more powerful with more energy and not less. That's the reason that the miracles back in biblical times will pale in comparison to the future miracles prophesied at the End of Days. God makes a statement through the prophet Zecharia in Zecharia 8:6, *"Just as in those days* [Messianic Times] *to the remnant* [the smaller number who are admitted into the Messianic Age] *of the nation, it will all seem incredible to Me* [God's miracles]." As part of those miracles, God will give us a new heart and a new spirit so that the energy of our soul will magnify in both quality and quantity. It is forecast that our children shall prophesize with this new influx of soul energy. If you think of our DNA, there are large portions of unknown functional DNA that do not code for proteins. God also has a large portion of energy that seems not to be used for the seven billion souls currently living on the earth. Perhaps God's untapped energy will be applied to perform the promised miracles in Messianic Days, including increased soul energy so that we shall be granted eternal life.

At the time of the big bang, when God's Light energy was there but was not released, it was essentially locked up. Noteworthy,

Stannard speaks about this locked-up form of energy as being associated with a mass at rest. In the moments before the big bang, when God's energy was released, it became kinetic energy. The difference between God and man is that God uses His released energy to create humankind through the soul entering the body; man uses his God-given energy for good purposes such as nuclear power stations, but also to create nuclear bombs meant for destruction of humanity. God will never let society as a whole be annihilated, as God is shaping the direction of the world whether we choose to believe it or not. He has a plan that will move us from the world of imperfection that He created beginning with Adam and Eve in the Garden of Eden to the perfection at the End of Days in the Messianic Age.

Our souls give us the special sense of "time" that Stannard refers to as the "Now," just like the soul creates our thoughts that happen so rapidly that we are instantaneously proceeding from one "Now" to the next "Now." We agree with Stannard that our consciousness is acting as a searchlight scanning progressively the time axis to label that special moment the "Now." We just attribute this ability, as we mentioned earlier, to our soul and soul consciousness, not to our human brain. You need the speed of the soul to accomplish the myriad of thoughts that come to us each day. The human brain is far too slow. Think of the human soul or the bodiless soul in the spirit world as whole and intact even when it appears to be fragmented. When laser light shines on a hologram, you get a three-dimensional image of an object. If you shine the laser on even a small part of the hologram object, the entire hologram image still appears. In this sense, a piece of soul energy still represents the entire soul like a droplet of water in the ocean represents the entire ocean.

What are the secrets of the soul? Does the soul have intelligence?

14

⌘

SOUL INTELLIGENCE

S EVERAL STUDIES DISCUSS SPIRITUAL INTELLIGENCE. Danah Zohar and Ian Marshall's book, *Spiritual Intelligence: The Ultimate Intelligence,* talks about SQ or the Spiritual Quotient, which is associated with parts of the brain that spur us on to live a meaningful life. Of course, other factors influence how a person behaves and how she sees the need to continually improve her character. PET scans indicate bundles of nerves in the cortex of the brain that light up when patients are queried on topics of spirituality. The spiritual quotient is distinctly different, therefore, from your IQ, which is your Intelligent Quotient, and your EQ, which is your Emotional (or feeling) Quotient. Your moral choices for good and evil are also part of your SQ.

Zohar and Marshall go on to attribute SQ to the soul. The soul thus has a brain but how this brain comes about has not yet been explained. They also talk about the character traits of someone who is considered to have a high spiritual intelligence, and the authors give strategies for improving your SQ, showing scientific evidence that neural growth can take place even at an older age. At the same time, we can lose neurons during the course of our lives. It is interesting to speculate that spiritual growth results in an increase in your SQ spiritual quotient. Does making ourselves whole make us not only better but smarter human beings? Are we wiser for it? Are we more creative?

Tony Buzan's *The Power of Spiritual Intelligence* contrasts what we readily believe and where we should be in our thinking. Most people view spirituality as something we need or do not need to squeeze into our busy lives. Buzan argues that we are not human beings having a spiritual experience. We are spiritual beings having a human experience. We can make that jump and agree with Buzan, but most people may not be able to do this. With so much enticement from our material world, we get exposed to an ongoing barrage of physically appealing influences. You can touch material objects that give you pleasure. You can't touch spirituality. You have to feel your faith by activating your spiritual sparks.

Spiritual intelligence is considered by many to be the most important among all the intelligences. However, we all know that you can live your life without spirituality, and you can be an atheist and be very smart. The spirit or "spiritus" means *breath* in Latin. Breath represents Mantak Chia's reference to *chi* or our life's vital energy. Without the spirit, or more correctly the soul, the body cannot live. When the soul departs at death, it realizes there is no longer hope for life in the body.

Many people often identify themselves by their material accomplishments; for example, how rich they are. However they die with their money while spiritual accomplishments live on in your soul. God gives you a lifetime to develop and grow your spiritual character.

Knowing where God came from, as we discussed earlier, is perhaps the most challenging question perhaps of all time. God has remained the most mysterious "person" even with atheists, who spend their time condemning the concept of a Supreme Being.

We can surmise that the soul is caring as during death in my past life regressions, the soul of David Samuels Smith and George Washington kept looking back. Losing the body is as difficult for the soul as we in the physical world feel in losing a love one. Souls shed tears and they watch over us from the spirit world.

"Does the soul have intelligence? Does the soul have a brain?" I spent about two months thinking about an answer to this question and then it hit me. Newton's patients who as souls visited Heaven in past lives-spirit world regressions were able to tell Dr. Newton under hypnosis what happened in the spirit world. They learned about their past lives and studied in libraries while spending time in rejuvenation centers. What caught my attention recently was the statements made about their souls entering the foetus between three months and birth. The key words were, "The soul over time completely integrates with the neurological circuits of the fetal and infant brain." I began to ask what does integration mean?

Checking with Marcia, I realized that each time you have a present life and then depart this world when the body dies, the soul's mass takes a miniscule neural piece of its own host human brain with it back to the spirit world. This implies that when the soul leaves the body in a particular lifetime, it does so with more brainpower than when it entered the body in that lifetime. The soul affects the body, and the body affects the soul.

For example, if you have lived for twenty past lives, you will each time accumulate brain neurons from your incarnated body, albeit tiny amounts. That means your intelligence fluctuates depending upon the quality and quantity of the neural circuits in each physical host from whom you essentially inherited a part of its brain intelligence. These tiny pieces of the brain along with your soul also helps you to recognize your past lives.

If the soul takes a tiny piece of the human brain in each past life, it stands to reason that the reverse is true and the human brain also retains a portion of the soul at least until death. This is based upon Newton's premise that there is a complete integration of the soul with the human brain. Newton also talked about the soul leaving for short periods of time, such as during human sleep to go out with fellow souls and explore familiar locales. How can this be if the soul gives life to the body? It's possible because even during sleep, the body has a small amount of soul to keep it alive.

I believe and Marcia has confirmed that her soul left her body seven hours before her death when I gave her permission to go and she shed a tear in acknowledgment of her agreement. She could do that as can any soul when death is inevitable in a dying body, because that dying body still has some semblance of the soul. At the time of death, that small piece of soul untangles itself and departs the body and joins the remainder of the soul that is close by. That is why death happens when the person is deemed to be "brain dead." They are never soul dead because the soul lives on eternally. It is also possible as suggested by Melinda Ribner in her book, *New Age Judaism*, that the *nefesh* part of the soul remains with the body when it dies. If enough soul energy stays with the body, even in death the body may not decompose to a skeleton.

In our eyes, intelligence of the soul isn't necessarily the means to wisdom or closeness to God. How you conduct yourself in a meaningful way in your earthly life, your spirituality, is what determines the closeness of your soul to God. It has more to do with the Light energy of your Divine soul and how you sublimate your evil inclination than it does with anything you will receive from your human brain or body.

Obviously God doesn't acquire neurons from a human brain. He has his own infinite organizing intelligence in His massive number of energy particles. Each new incarnation brings a new piece of God to the Divine soul with a special mass of energy that possesses a piece of God's intelligence. Intelligence of a soul is thus what God has given you, plus the contribution of your human host. Our judgment is that such a process makes no two people alike and can create bright unique individuals like an Einstein. My Marcia is highly intelligent, and from energy measurements, both of us belong to a soul group closest to God. When I had my past life-spiritual regression, I visualized a brilliant white yellow, almost white-gold cluster that blinded my eyes. It was so bright that it was like looking right into the sun. I had to close my eyes several times under hypnosis.

We said before that the soul inside the body is constrained. Marcia's soul in the spirit world is free, and she can read my thoughts while I cannot read hers. Her soul is one hundred times more potent than mine. The internalization by the soul of the body's neural circuitry is what gives the soul the ability to retain the five senses. However, some of the senses are available to the soul only at a reduced level.

The soul has touch but it's a light touch. Marcia has visited me in our home and poked me when I was falling asleep on the couch. When I had my hypnotic regression, Marcia and I were hugging and kissing, but it wasn't like the earthly hugs and kisses we would passionately have in the physical world. Also, I could not smell the orgasmic breath coming out of Marcia's nose that I longed for when I kissed her.

Souls are happy in the soul world where they can sing and dance or just be at peace with themselves. The spirit world has no stress. Souls can use their powerful energy to create their human form and they see and hear you. They can smell flowers and fragrances, and they can smile, but they don't laugh. Laughter requires more neuromuscular activity that the body supplies. Souls do not eat for the same reasons, but they do rest and sleep according to our energy determinations.

Souls appear at half-mast transparency from the waist up. Depending on the soul, the soul can place its energy into earthly animals such as butterflies, hummingbirds, dragonflies, and squirrels. In Marcia's case, she gave the yellow and black monarch butterfly unique cyan blue and red circles and Roberta's squirrel a unique orange-red crest. How souls do this is a total mystery. The same holds for Marcia visiting our home and joining me for a movie on television. She remains in Heaven, but she can project her energy and is invisible to me while she is watching TV. Marcia never saw the *Bourne* movies until she became a soul in the spirit world, but she can now tell you everything about the movies she watched with me. "They were good movies, Jerry."

It is appropriate next to further try and understand human and soul function as both independent and coexisting entities.

15

❧

SOUL & HUMAN BRAIN
PHYSIOLOGICAL FUNCTION

NIMALS WITH LIMITED EXCEPTIONS DO NOT HAVE LANGUAGE.
It appears to be a gift from God to humans. This may
not be a surprise because almost four thousand years ago,
God confused those building the Tower of Babel in the Holy
Land by giving them different languages so that they could not
communicate with one another. In modern times we might expect
language to originate in the human brain, but it does not. It
comes from the energy of the Divine soul. Unlike thought, which
similarly originates in the soul, language is expressed through
the human brain, whereas thought is not.

I find it intriguing that if you listen to a tape of your voice,
it doesn't sound like the voice you hear when you speak aloud.
I sound so different listening to my voice than I sound if I tape
myself and then listen to the tape. Most people have a unique
voice that they project to others. Where does this unique voice
originate? Marcia, please answer.

"It's from your Divine soul, Jerry. People hear your voice that
originates in your soul and is then expressed in your human brain
by the saliva, tongue, larynx, and voice box. The combination

of the two, your soul and human brain, gives each person that unique voice that projects out to other people."

"But why do I hear a different voice when I speak?"

Marcia answers and I check her answer through my energy measurements. "You listen to your own voice only through the speech apparatus of your human brain. Your soul is not involved." She elaborates further, "An entertainer can mimic the voices of other people, but this requires energy, the energy from his Divine soul."

I then asked, "What about you? You are a soul without a body. Why do you sound exactly the same as you did when you were alive on Earth?"

"Because, Jerry, as we wrote in the previous chapter, I took a piece of my human brain with me when I departed this Earth last year. I still have all my five senses and an ability to speak to you telepathically."

"Can you speak to me aloud?"

"I could, but we are far apart and you would have to hear me by sound waves, which are much too slow. To speak to you aloud, I would have to use most of my soul energy. Right now we communicate telepathically so that our thoughts are carried by the speed of God's Light requiring only a small percentage of my energy. You can speak to me aloud because you have all of your human brain function. I don't"

"Oh, I see. What about God?"

Marcia responds, "God is different than all others. His energy is limitless and so He can speak to you aloud as He did to you back in 1982, when you were living alone in your cottage in Poquott on Long Island. God was able to speak in a booming powerful voice to the thousands of Israelites at Mount Sinai. The huge numbers of Israelites necessitated God speaking this way, and He wanted to make a lasting impression. After He delivered the first two commandments, the Israelites were so tone weary that they asked Moses to deliver the next eight commandments, which were passed to Moses by God. When you heard God in Poquott or Erin heard Him during a walk near our New York house, or when Moses or Jacob heard Him one on One back in biblical days, God spoke to all of you in a quiet assertive tone. God's voice is also unique. You heard Him in 1982 and then again at the beginning of 1999, believing God spoke in one and the same voice. Back in 1995 when I was contemplating splitting up with you and went into the bedroom while you stayed close to our living room by the front door, God spoke to me in this same quiet assertive tone."

"Where do wisdom and imagination or fantasy come from, Marcia?" My energy measurements tell me that they originate from the Divine soul and not in the human brain.

"You are correct, Jerry."

"Then wisdom is different from intelligence," I said. "Wisdom is like your spiritual intelligence, SQ, which originates in the soul, but your IQ, or rational intelligence, and your EQ, or emotional intelligence, originates in the human brain. Your SQ helps you improve on your EQ, your feelings about yourself, and perhaps some on your IQ, your knowledge and logic. Am I correct?"

"You are right in all what you say, Jerry. Both your spiritual intelligence and your wisdom originates in your soul but are expressed through your human brain, like your voice. Fantasy, imagination, and love of art and music originate in your soul but are similarly expressed through your right brain, but spiritual intelligence and wisdom connect with both your right and left brains. The soul eventually totally integrates with all of your human brain and your eyes after it enters the fetus."

"Then is there any difference between your spiritual intelligence and wisdom? King Solomon was thought to be the wisest man alive. Einstein, on the other hand, was perhaps the most imaginative and had a good SQ, along with a brilliant IQ and special EQ." I checked my energy and I got a Yes on the difference.

Marcia explained. "Wisdom is all your years of accumulation of spiritual intelligence throughout your life. The more you climb the spiritual ladder, the more you increase your spiritual intelligence and your wisdom. Job in the Bible said it best, '*Days speak and a multitude of years teach wisdom.*'"

I moved to another topic. "What about doubt?" I asked. "God gave us doubt so that those disinterested in Him could exert their free will and not believe in Him."

"It's the left brain which also controls procrastination, Jerry. There is no connection to your soul. The same is true when you dwell upon and fear death."

"And love, what about love?" Marcia knew about love. It was what she dreamed about even at age twelve, finding someone to share her life.

"Love is different," she said. "True soul mates or true love originates in the *pintele yid* in your heart and not the human brain or your soul located in the brain."

I said, "That makes sense. The *pintele yid* is that part of your spiritual soul that God hides, and you need to spark it when you begin your journey climbing the spiritual ladder. Is the love expressed through your brain?" I asked.

"No," Marcia said. "It originates in the *pintele yid* part of your soul and is expressed through your physical body heart. The heart has numerous neurons travelling through it, like the brain. Happiness, joy, laughter, and sadness are also part of your soul heart and are expressions of both your physical heart and your human brain. You feel rejection and sadness in your heart, like the expression, 'a broken heart.' Your human brain is responsible for your dwelling on this sadness or putting a spin on emotional outcomes. Grief is the same as sadness, like you feel for me, Jerry. I don't have a physical heart, but my soul free of my body still feels sadness and love for you."

"And the negative emotions like anger and jealousy?" I inquired.

Marcia replied, "You already know the answer to this question. These emotions have nothing to do with the soul, for these emotions are solely felt in your left human brain. The evil inclination influences you to feel these emotions, but sometimes anger is justified, although jealousy never is. Your spiritual soul helps you to eliminate jealousy if the human part of you is envious and jealous. You realize that everyone on this Earth is significant in his own way. Jerry, you released your jealousy by becoming spiritual." Jealousy was expressed as an undesirable trait by the tenth commandment.

"What about anxiety, terror, and panic?"

"These emotions, as you know from being a scientist and having undergone many years of primal therapy, come from your lower brain, your salamander brain. They are very hard to eliminate. Primal therapy helped you, but so did your spirituality. Anxiety is unfocused fear, meaning you are not aware of where it came from because you are not in touch with your feelings in the womb unless you regress back in time and do therapy."

"It was once explained to me by my therapist. If you are standing on the sidewalk next to a busy road and a tractor-trailer truck comes by, your anxiety will kick in and you will believe that the truck will hit you, just like you almost died during the birth process. Most people would not make the connection of reliving their life-threatening birth and would not realize that the anxiety was coming from the womb. However, if you don't feel anxious when the truck comes by, you might still feel fear in the present because of the size of the truck. You just won't believe that the truck will hit you. Fear may always be with you. It's a natural feeling response. After you feel your fear, you can then overcome it and act on your adrenaline like a person performing a heroic act."

I continued. "I'm sure faith is strictly within your Divine soul in your brain because it is within the domain of God?"

"Yes," Marcia responded, "but it is also initially located in the soul energy of your *pintele yid,* which you need to spark if God has not intervened in your life like He did with you, Jerry. After you activate your *pintele yid,* faith and belief are now part of your physical heart as well as your Divine soul. Your passion comes from your physical heart and also your human brain; but if you are feeling passion about God and spirituality, your passion also comes from your Divine soul."

When I communicate telepathically with Marcia in the soul world, I don't see that passion in her. It's there to a small degree, but it's not anywhere what it was when she was in the physical world, because passion requires your physical self. She does tell me in the same way that she loves me, but it's not passionate. According to Newton, souls may have a different kind of energy passion when they sing and dance together.

Your type A personality comes from your brain, as does seeking wealth, unless you have good intentions to use that money for spiritual purposes. Then your soul kicks in to help you with your philanthropy. Being a philanthropist doesn't necessarily guarantee that the individual will not have faults. God created us imperfect, so we all have faults, except my Marcia, of course. Marcia just told me that she doesn't blush as a soul, nor does she get embarrassed. More of your brain is required for these actions.

The functions of the human brain and the soul seem to be a mixed bag. Sometimes they work together, sometimes they perform independently, and sometimes there is an expression in one entity after originating in the other entity. Perhaps what's important is that we are highlighting the soul as a distinct entity playing a significant role in our lives and our afterlives.

Orthodox Judaism talks about the energy of Divine soul as not occupying space in the body. Einstein and other physicists would tell you that all space is occupied by something. So how can the soul exist if it doesn't occupy space? It may be that soul energy as it travels in the body can be detected at chakra points but because the energy is continuously moving, we can't find it in any one place in the body where we would be able to weigh it. According to Einstein, all energy has mass and all mass has energy. We can remove blood from the body and observe and weigh it in the clinical or research laboratory but we cannot remove and bottle energy, at least not yet. God's will share his secrets of His energy in the coming Messianic Age.

Where do Past Lives fit in to the soul story? Religion seems to be meaningless if we are Jewish today and Christian one-hundred years ago.

16

PAST LIVES

E RECOGNIZE OUR PAST LIVES BECAUSE WE ARE CARRYING A TINY AMOUNT OF ALL PAST SOULS IN OUR NEW SOUL. We are also carrying information from the human brains of our past souls, and our brains carry bits of our souls. Einstein died in 1995 and his soul went back to the spirit world. According to Newton's studies, Einstein's soul likely incarnated into a new body. Because someone's new body has a piece of Einstein's brain, there is probably a budding genius trying to give us a unifying deterministic theory of the universe. Einstein loved Mozart and played a serious classical violin, so perhaps this new descendant is a rock musician. Maybe if souls are shared by family members, this is how we might explain how creativity comes out in different ways. Einstein's brain was taken from his body, and the culprit who did it shipped parts of his brain all over the world. He did not know that a part of Einstein's brain survived and is alive and well.

In my first past life-spirit world regression with Jules, I met up with Moses and found myself carrying his staff up the mountain. I was the soul of David Samuels Smith, but it is possible that I am also carrying Moses's soul. Wow! What an ego trip. What's interesting, when I was writing my book *Messiah Interviews* back in 2008, is that I used a fictional Time Line Therapy to get the protagonist to Heaven. Before I met Marcia, I was out to dinner with my friend Howie in 1982. Howie told me about Time Line Therapy. Here is the story.

It was a warm Saturday night, and we had finished our hamburgers and onion rings at the Good Steer restaurant in Lake Grove, Long Island, New York, when I suggested we go back to my office at Stony Brook University. I turned the lock of the outside door, and we surreptitiously entered Building L, Westchester Hall. No one was about outside or apparently inside. The building was deserted as all the expert-thinking scientific minds had taken their thoughts home for the weekend. We sat in my ten-by-ten cubicle of an office and turned off the fluorescent lights. It was a warm, hazy, moonlit night, and I had had some red wine, probably too much, with dinner. I felt unusually relaxed and lightheaded. Even thoughts of my research experiments had faded, which was a rare occurrence. Wherever I was or whatever I was doing, I would be thinking science. Science was my life, my first love. I was going to win the Nobel Prize one day. I told Howie that I was all ears and ready to focus.

Howie: Imagine you are parked on a deserted street. Not a soul is around, and you are sitting behind the driver's wheel looking up at the stars and the moonlit night. Now just breathe as slowly and deeply as you can.

Jerry: Shall I remove my glasses?

Howie: Yes. Try to relax, and imagine leaving yourself in the car as you are now, but also take yourself across the street and sit down on the curb. Look at yourself in the car, from where you are sitting on the curb.

Jerry: Is the me still in the car supposed to know that I am being observed by myself?

Howie: {Exasperated} Forget about the technical stuff. Just concentrate. If you want to both self-observe and self-remember, then fine. Let the heaviness depart from your body and try to sink down into a meditative state. Let all the tension in your body sink through your feet into the floorboard of the car. Permit yourself to enjoy this happy feeling. Let all your muscles relax.

I had meditated on several occasions, so I was confident I knew how to do this. I felt my head tilting slightly to my left side, and I was freely secreting saliva from my parotid and submandibular-sublingual salivary glands. I started to drool, a familiar physiological characteristic of my altered meditative state.

Howie: Now see yourself becoming light and shapeless, and freely floating out the window. You are flying upwards with absolutely no effort. You are as light as a feather. Let yourself fly. Let yourself go. Let your whole being rise to Heaven.

And I allowed myself. I flew upwards, higher and higher. I landed upright on a mountain, like Superman would land, and then found myself descending, carrying the tablets of the Ten Commandments. I was Moses. I even looked like him, long white beard and robe. Standing at the foot of the mountain, just waiting to pounce upon me, were Israelites with hostile eyes. They were blocking my way, and I feared for my life. I began to move through the unyielding crowd and they reluctantly gave way. It was as if I was walking an Indian gauntlet, trying to keep my composure, hold my breath, and simultaneously disguise my fear. I felt that they were ready to murder me. Then, in the distance I saw two smiling adult women. My faltering courage was fortified, and I felt a huge sigh of relief. I recognized my daughter Erin, in the present only nine years old. The other girl had blondish-brown hair and was much taller than Erin, but I couldn't identify her at the time. I think I know today that the taller girl was my granddaughter, Sarah, all grown up. Both girls were biblically dressed like everyone else in this surreal scene. The two women wore babushkas or kerchiefs around their heads, like I remember both my grandmothers wearing. I pondered momentarily. Were

they my wives in a past life, and not my daughter and granddaughter?
My eyes opened.

I had a smile on my face and a feeling in my heart and soul of inner
peace. When I described what I had experienced to Howie, he remained
silent, trying to digest what I had experienced. I don't think even he
expected a Moses time line regression from me. How could I convince Howie
if I couldn't convince myself that what had just transpired was true and
that, indeed, I had been Moses in a past life. I tried to change the course
of our conversation and said, "I don't think it would be unreasonable
that Moses would be frightened. After all, the Israelites had grown
impatient while Moses had been on top of Mount Sinai. They had built
the golden calf." Yet I wanted desperately to believe that I had been Moses
in another life.

In 1983, when I returned to Israel on my honeymoon with
Marcia, the highlight of our trip was climbing Mount Sinai.
We carried the *challah,* and at the peak I chanted the Hebrew
prayer blessing over the bread. I kept waiting for something
Divine to happen; however, it never did. I was disappointed. We
both laughed as we descended the mountain and caught up with
our small group at St. Catherine's Monastery. We had travelled
by truck through the desert *wadis* and returned to Israel's Ellat
after a border check on the Egyptian side. Our Bedouin guide,
Achmed, bid us good-bye, and I slipped him a tip. I also gave
forty dollars American to our Egyptian guide, Jimmy, and asked
him to buy us a painting in Cairo and send it to us in New York.
The other Egyptian on the trip gave me a sideways kind of look
that meant, "Are you a naive stupid jerk?" I scribbled down my
address on a scrap piece of paper and of course never heard from
Jimmy again.

Newton touched on the people you knew in your present life
that appeared in your past life. Back in 1982 in my time line
regression, when I was Moses carrying the tablets down Mount
Sinai, my two wives were my daughter Erin, who was only nine

years old back in 1982, and another woman who had lighter hair. The latter wife whom I thought was my granddaughter wasn't born until 1998, so I was having a past vision into the future. Checking my energy now and confirming my thoughts with Marcia, my wives back then, thirty-three hundred years ago, were indeed my granddaughter and my daughter Erin. However, because I didn't have a face to put on my granddaughter, the woman I saw did not have my granddaughter's appearance although there was a resemblance. We have previously explained this visual phenomenon when we talked about Martha Washington in my second past life-spirit world hypnosis regression. Two important points to remember are that we carry 1 % of the souls of each past life and that the 1 % can be thought of as the entire soul of the person that you visited in your past life. This implies that you become that person in the past life while under hypnosis. You may speak a foreign language in that past life but when you return to your own body in the present after the hypnosis is over, you can no longer speak that foreign language.

Where was Marcia in all this? She was Moses's sister Miriam. This too is common in Newton's past lives research where my soul mate in a past life might be my mother or son or daughter, or in this case my sister. Miriam was the older sister who watched over Moses as a baby when the pharaoh had ordered the murder of all Hebrew newborns. The pharaoh had taken note of the increasing numbers of Hebrew slaves and he felt threatened.

Miriam was known for being a leader of women, and she was a beacon of hope for the Israelites during their Exodus from Egypt. She was also married to Calev, a descendant of Judah, who was one of Jacob's twelve sons and one of the Twelve Tribes of Israel.

Miriam's marriage to Calev is significant because twenty-five hundred years later, a descendant of Calev and Miriam was David, who became the second king of Israel. It is from King David's lineage that the Messiah will arise according to the Hebrew Bible.

In the New Testament, Christianity claims that this is Jesus. I mentioned earlier that I believe that I am carrying 1 percent of the soul of Jesus. What I didn't mention was the amazing feeling that I had that I couldn't understand during my PhD studies in Israel from 1966 to 1969. Wherever I walked where Jesus had been like on the Via Dolorosa in Jerusalem or the Sea of Galilee where Jesus gave the Sermon on the Mount or in Nazareth where he rode his donkey, I felt this awesome feeling that spread through my entire body. I feel it now as I am writing these words. Judaism believes that the Messiah is someone currently unknown to the world. We could bring Christianity and Judaism together if we postulated that the Messiah is carrying 1 percent each of the souls of Jesus and king David. The Messianic Age is not the age of the Messiah, but is really the age of God. It is a time of peace, harmony, and love, which seems not to be destined in this world because of man's inhumanity to man. No one knows how the Messianic Age will come about. In the prophetic descriptions spouted in both the Hebrew Bible and the New Testament, the ending for our world is apocalyptic, full of fire and brimstone. However, God is the ultimate miracle worker, and it shall be His Will that shall prevail.

Science cannot explain miracles. Neither can miracles explain science.

17

DIVINE MIRACLES

I DON"T THINK I WOULD BE ALIVE WITHOUT DIVINE PROVIDENCE COMING INTO BOTH MY LIFE AND MARCIA'S LIFE. I know this for a fact. God came to Marcia and she saved my life when by all rights I should have been dead after my suicide attempt in March of 1995. Marcia gave up her life for me so that I would come out of my recalcitrant agitated depression. I have been blessed by God, and for many years I never understood why. I have some inkling now that this incarnation of souls into new bodies is to be the last one, and that we will soon enter the Messianic Age. Marcia will not therefore go into another body, but her Divine soul will go into her dead body lying under the ground and resurrect her. It is my belief that Marcia and I and our family are connected in some way to the advancement of the Messianic Age. That is our mission and that is why we come now across two worlds to write this book. The story is only half a story unless I describe how this sinner who is me was blessed by God.

By the fall of 1982, I wasn't yet married to Marcia and was living in a cottage on Long Island, separated from my first wife and my three children, Melanie, Seth, and Sean. I was promoted at age forty-one to professor in the oral biology and pathology department at Stony Brook University and had begun primal therapy. It had been a year since I began my therapy, and it was clear that I wasn't making progress. I had filled up seven research laboratory notebooks of feelings, each book consisting of three hundred pages. There was a lot of repetition, and I seemed to have reached a plateau in terms of making further progress. In an unusual display of frustration, I brought the seven notebooks to the edge of my Poquott loft, and proceeded to throw all of them in one thrust onto the floor below. I felt like Moses smashing the first set of tablets on which the Ten Commandments were inscribed. The sounds of the books crashing were silenced by a loud and strong male voice that was definitely external. I heard the words very clearly, "And you shall be Mine."

I remember being both very startled and very afraid, because I neither knew whose voice this was, nor could I figure out how the voice got into my cottage. There was no bipolar disorder or manic depression in 1982, so I couldn't blame the voice on mental illness. I heard the voice again, about two weeks later. This time the words were different. "And you shall have." Could this be the voice of God? I thought so ten years later when I revealed this mysterious voice occurrence to my Christian therapist. She believed me without a shadow of a doubt. I went on with my life, but the memory of the voice never left me.

Why did this happen? I had no explanation. I kept the notebooks recording my primal feelings for many additional years before angrily dumping them in the garbage along with my primal therapy textbooks. I was to discover twenty-five years later that I would need these books again when I returned to face my demonic maker, my mother, once again in the womb.

During my womb regressions, I vividly felt that I had been saved by a miracle during an abortion attempt that took my twin brother's life. When I was being sucked up along with my twin, a force pushed me in the opposite direction and held me there until all was quiet. One night while I was wide awake in my bed reflecting upon my prenatal experiences. I felt myself gliding upward, levitating under the control of an untouchable force of energy that totally surrounded me. Perhaps it was my soul rising, as I now reflect back on the moment. Yet I was in a completely non-meditative overtly conscious state, and I could not have been accessing my unconscious like I recently did under hypnosis. However the feeling of gliding upwards was the same. I found myself being carried upward while suspended as if I were floating in water. I thought of my description of myself to my therapist when I was floating as a foetus in the amniotic sac of my mother's womb.

As I travelled upward, my eyes gazed upon a passing solid shape that came into view. I clearly saw segmented stones, similar to what my ex-wife and I had visualized in the Western Wall on our Israeli bus tour the day after the Six-Day War. However, it wasn't 1967. It was the tail end of 2006 and this stone wall was under water. My levitation gently came to a halt.

I could feel stares. It seemed like I was being examined by a group of formless *creatures* and a man with a full beard, who was definitely in charge. Were these the stares of angels who had abducted me, like the stories you hear about Martians or UFOs? The formless creatures were pensively swaying from side to side. They seemed to be in deep thought, trying to come to a decision. "Could this be the one?" They repeated, "Could this be the one?" And once more. I still have the head man's face in my memory. I can see Him vividly in my mind. "Was this a man?" I thought. "Was this the transcendental God?" I was then lowered back down onto my mattress. It was all over in a matter of minutes. I remain perplexed.

In 1991, I had my first bipolar disorder episode. I became manic, and then I went into clinical depression. I was hospitalized at eastern Long Island's mental health Greenport facility and received electroconvulsive or electroshock treatments. The next few years were touch and go. In late 1994, I had a second major breakdown of agitated clinical depression, which caused further hospitalizations, and finally forced me to take a five-year long-term leave of absence from Stony Brook University. I had always maintained that the worst part of being in a psychiatric ward was when the door locked you inside with no exit to the outside world.

My freedom had been taken away from me. And as strange as it now seems, my greatest fear as a child about being locked up forever in 999 Queen Street, the insane asylum in Toronto, had symbolically come true a half-century later. I was now a card-carrying member of the mentally ill, and at the time I thought my destiny had been sealed for the remainder of my days. There would be no return to so-called *normal* life.

It was during this five-year mentally ill hiatus from work that I experienced miracles, which I wrote about and published in my books, *Divinely Inspired* and *Messiah Interviews*. It was also in March of 1995 when my agitated depression became unbearable, and I attempted suicide. I was almost successful, but once again was saved by Divine Providence. A few minutes more without help, and my life would have been over. Help came from Nesconset volunteer firemen, who rushed me to the hospital. It also came from God, Who implanted a supernatural message in Marcia's brain for her and Erin to return home.

They were forty-five minutes away and had just ordered lunch in a village diner when an extrasensory perception from an unknown source revealed to Marcia that something was terribly wrong with me. They found me lying in bed, drifting off peacefully to greet death and my Maker. I seemed content and I appeared calmer than I had been for a long time. The thought of losing me was unbearable to Marcia. Besides, I guess it wasn't my time. Marcia jarred me hard and Erin called 911. The fire department responded, and the doctors in the hospital saved my life on that cold day. It would take another six months of agony for me to get better. This almost certain-death episode was followed by a much milder suicide attempt and further psychiatric hospitalizations. After Marcia told me she wanted to leave me and then changed her mind, it was two months later that the clinical depression lifted as suddenly as it came. Medication had no role in my recovery. Was it Divine intervention? Was God once again there to help me? I think Yes. If Yes, why? Why so much help from the Creator on my behalf?

The miraculous events that transpired beginning at the end of 1998 changed my life and brought complete wellness. I had no further bipolar disorder episodes, and I have been in remission until the present time. I had recovered and went back to work in September of 2000, and Marcia and I began to make plans to retire. We set a time frame of five years. During the Christmas break of 1998, we drove down to Miami Beach and parked ourselves at the Fontainebleau Hotel. It was a beautiful sunny day in South Florida. The clouds in Florida are wonderfully unique. They are so magnificently angelic white and puffy when set against a powder-blue sky.

We were lying on our lounge chairs and Marcia was sleeping. For some reason I looked heavenwards. What I saw was perhaps the most incredible sight I have ever seen. Pictures of myself as a little boy, a teenager, a young adult, and then an old adult were flashing in the sky in rapid succession, just seconds apart. Quite remarkable was the fact that these heavenly photos were the same ones that I had stored away in my picture album.

I knew this to be true, because I often looked at these unhappy pictures. There were one or two photographs of myself as a cute, blond, curly-haired, smiling little boy that I loved. I was forever during my life trying to recapture this happy little boy within myself. Unfortunately the identical photos in both my albums and in the sky also showed me as an ugly and neurotic teenager with tightness and tension lines in my unhappy face. It was only by luck and a phony smile that I was able to pose for any kind of camera picture that I could live with. One of the main reasons I told my therapist why I entered primal therapy was because I couldn't smile.

I had started to lose my already thin hair in my early thirties and amazingly this loss of hair was depicted in the sky. From mild loss to the full hair loss horseshoe pattern, I clearly visualized it all directly above as heavenly flashes. Not only did I see myself at my present age, but also I was being given an extraordinary glimpse of myself in the future. There were several repeated remarkable flashings of me well into old age. The remaining hair on both sides of my bald scalp appeared almost wildly Einsteinium as I aged. But for all this time since 1998, I couldn't understand why, because my hair never stuck out like Einstein. As I talk now about my hair, I realize this look was not a coincidence, because thirteen years later God knew that Marcia and I would write this book, *Putting God into Einstein's Equations*.

I wore a hairpiece, but God saw and loved me in my natural state. There was no hairpiece in the sky. After the slide show ended, I was numb with wonder. It seemed as if God was telling me that I would live a long life. But what about Marcia, who was asleep peacefully in a lounge chair all this time?

All of the miracles I experienced became the catalyst for my biblical self-study. I wanted to divorce myself from my sinful past, refine my character, and embark on a new spiritual road. I had a thirst to seek a higher purpose. I immersed myself in the Hebrew Bible, God, evolution and Creation, human anthropology, and a host of other subjects. I read voraciously, waking up in the wee hours of the night and then discussing everything in inquisitive back-and-forth one on one Talmudic fashion with Marcia. I tried unsuccessfully to come up with reasons for free will and the existence of humankind with its good and evil inseparability.

Marcia became my conscience, and we continued our discussions well into the night until we became so exhausted that our brainwaves stopped recording. It was during this period that I developed new spiritual thoughts about our purpose on this Earth and in the World to Come.

During our vacation in Florida at the end of 1998, I saw a beautiful white egret with an enchanting slender neck fly directly past the front of our car. Moments later as Marcia dozed, I saw cloud-like white letters in the light-blue sky, spelling out the name of my son Sean with the *S* capitalized, followed by the *ean* in lowercase letters. Within a fraction of a second, the *an* became *th*, now spelling out his identical twin brother's name, Seth. I was dazzled. My eyes remained glued to the sky. I saw my sons' names, Sean and Seth, rapidly alternating for a full ten to fifteen seconds. Immediately after this, I saw blurred writing in the sky. The first capital letter, *M*, was absolutely clear, as were the small letters, *ia*, at the end of the word. However, I couldn't make out the middle of the word, no matter how hard I tried straining my eyes. It all happened so fast.

When the heavenly visions had disappeared, I started to think about the blurred word and realized that perhaps the answer could be found in the initials of the first names of family members. My wife's name, Marcia, began with *M* and ended in *ia*. But the family also had the letter *M* from our eldest daughter, Melanie, and my son-in-law Magnus as well as my daughter-in-law Karen's middle initial. I was being sent a message in the clouds, and the word was blurred because there were three family members whose names begin with the letter *M*. Only Marcia's name correlated with the *ia* at the end of the blurred word. The pieces of the puzzle were beginning to fit together.

I believed that the naming of our five children had been random. Marcia was my wife from a second marriage, and my biological children were Melanie, Seth, and Sean. Her biological children were Erin Michelle and Kenneth Scott. Kenny had married Karen Michelle, and the two grandchildren were named Sarah and Ethan. In a remarkable insight, I realized that the first letters of the given names of our children and grandchildren corresponded to the first four letters, "Mess" of the word *Messiah*. E was for Erin or Ethan, and *S* and *S* was for Seth, Sean, and Sarah and was Kenny's middle initial.

I could fill in the "ia" of *Messiah* from the ending *ia* of the blurred word in the sky. Also with regard to the *i*, I realized years later that in Hebrew, the *I, J,* and *Y* are interchangeable, and therefore the *i* of "ia" could also represent Marcia's parents, Irving and Jean; my father, Jack; and me, Jerry Joseph. My mother's name was Anne, so that would take care of the "a" of "ia." The final "h" I could find in my granddaughter's name, Sarah. *H* is also the first letter of my father's Jewish name, Haikel or Chaikel, like Hanukkah or Chanukah. Was this a coincidence, or was this Divine Providence? Our family seemed to be in some way connected to the Messiah through our names, but how?

Interestingly the surname Polak, my paternal grandfather's actual last name when he was born in the Ukraine, is a surname found on one of the three groups of families that trace their family tree back to king David. Very recently I discovered a DNA group, the *Yicchus* group, of these surnames. Upon checking my own DNA of my Y chromosome against these individuals thought to be on king David's genetic lineage, I was an exact match in all 67 genetic markers. In addition to a family tree and DNA lineage, there is thus, as we have illustrated, an independent soul lineage that doesn't have to follow these two other lineages because God assigns the past souls that we carry. Am I and my family members carrying 1 percent of king David's soul? If so then this would explain our connection to the Messsianic Age.

While still in Florida, I saw the name *Ruth* in the cloud-writing in the sky. When we got back to New York, I read my first biblical book, the book of Ruth. Another coincidence, perhaps, but Ruth was the great-grandmother of King David. In that first week back in New York in January of 1999, I experienced more unusual happenings, such as hearing God's voice a third time. It was seventeen years earlier that I had heard God's voice in my Poquott cottage. You can read about all of this in my books, *Divinely Inspired* and the *Messiah Interviews*. Then in August of 2000 upon returning by car from Florida, I saw my first ray of blue light coming through the side window of the car. Days later I saw the blue ray between two normal white rays of the sun. Then the rays of light were reversed, and the white ray was enclosed by two blue rays. These spiritual lights continued to flash in front of me for two years more until I thought it was time to write my first book, *Divinely Inspired*.

The last miracle took place in March of 2011 under very sad circumstances. It was the forewarning of Marcia's impending death. Erin, Marcia, and I returned from the H. Lee Moffitt Cancer Center in Tampa, where we had scheduled a procedure for the following week. When we arrived back at our home in Boynton Beach, I went into the office and saw it. Saw what? Suspended in midair was a fistful of smoke with its distinct burning smell. I immediately thought this was our miracle. God was going to cure Marcia of her cancer. I was completely off base as my nephew Scott explained that last night before Marcia died. God was warning me that I could not save Marcia by my plans to get her cured through medical science treatment in clinics in southern Germany and Vienna. God was telling me that He was taking her back from me. It was Marcia's time to pass.

How does God create miracles? We know how massive His energy particles are and how we pale in comparison to Him as Divine souls. How could He split the Sea of Reeds during the Exodus, and how can water run upward during the separation of the waters when water's natural tendency is to flow downward, a miracle within a miracle? How is it possible for Him to be everywhere monitoring our moral actions, and how does He know absolutely everything that's going on in the world without directly participating?

To carry this a step further, how did God create Adam from the stardust from the four corners of the Earth? Adam was not born out of the union of a sperm and a seed. He amazingly appeared as a handsome youth and was transported to the Garden of Eden. And when Adam was lonely, God became the first geneticist and took Adam's rib and cloned Eve to be his woman, Adam's soul mate. Wasn't this also a miracle within a miracle? Eve's DNA should have been identical to Adam's, but it wasn't thank God, for the continuation of our human race and the splendor of sexual harmony. Adam and Eve should have been identical twins, but male-female identical twins are not within the realm of scientific reality. God seems to be the greatest of all superheroes for all eternity.

I posed this question of miraculous miracles to Marcia. "It's all energy, my love."

"But how?" I pressed. "Do you know?"

"Yes," she said. "Think about Einstein and his theory of relativity, Jerry. Energy possesses mass and mass possesses energy. We as souls can appear as animals like the butterfly or the squirrel because we can use our energy, especially being free of the body, to do this just like I can show my soul in transparent form to you in the spirit world or on earth."

"But," I argued, "the animals that we saw like the butterfly have solid physical mass and you don't."

"You are quite correct, Jerry. Our energy actually fuses with animals, and the mass within our energy takes on the shape of the animal. Therefore, we create your butterfly or squirrel with their unusual markings. Don't forget that we are light and light has the color spectrum, so we can add blue and red dots to your butterfly or give the squirrel an orange-red crest."

"Is it the same butterfly that I see month after month at your gravesite?"

"No, Jerry, the butterfly has a short lifespan. The more advanced souls can pick another butterfly and recreate the red and blue dot markings like the graphic designer does on his computer."

"From everything you tell me, Marcia, you should be able to do the same with physical humans."

"We are not allowed, my love. God forbids such human modification. This only happens at the time of incarnation when our soul energy goes into a new foetus."

Marcia continued, "The Devil, however, has no qualms about using his dark energy to incarnate into a foetus or existing human bodies to create evil demons and devil incarnates. In a sense, the Devil has descendants as you wrote in your book, *Gog and Magog: The Devil's Descendants*. That was your best book, Jerry. I know you know about evil because you have had contact and discussions with the Devil."

"Yes", I said, "the Devil and I have choice words for each other. We have spoken often these past months as he interferes with my thoughts. He even speaks exactly like you, Marcia, so I have to constantly check my energy to see who is speaking, you or the Devil."

"I can use my energy to try and block him out, Jerry, but it's very difficult. The Devil is very powerful and probably only God can stop him."

"OK," I said, "I see how all this works, but what about God's miracles?"

"I don't know the answer to that question. I suspect it is related to the Creator's limitless omnipotent energy where He actually can create humans like Adam and Eve from scratch or alter nature at His Will, like showing the burning bush to Moses or sending an unending supply of quail to the ancient Israelites when they were complaining in the Sinai Desert. It's a well-known trait that Jewish people know how to complain and give their opinion. God can also act through individuals like Moses or Jesus or Joshua who miraculously made the sun stand still."

"I guess God is allowed to have some secrets," I said. "The prophecies state that at the time of the Messianic Age, we shall know God as the waters of the sea."

"Yes, my love," Marcia added, "God will explain all at that future time to humankind when those admitted into the second Garden of Eden shall live forever."

In 2001 a thought came into my head from out of the blue that I should start a non-profit charitable corporation for the Third Temple, which would be essential to the coming Messianic Age. The Third Temple was the Creator's promised dwelling place on Earth. It was where His *Shechinah* or Divine Presence is prophesied to reside at the End of Days, as it did in the tents of the Patriarchs, the Tabernacle in the Sinai Desert, and in the First Temple centuries ago. I gained non-profit status by successfully obtaining a 501(c)(3) classification for the Shechinah Third Temple, Inc. The Temple will likely be built by the Messiah after the prophesied great earthquake, which will occur during the wars of *Gog and Magog*.

I have never been a practicing religious orthodox Jew. However, my steadfast belief in the Messianic Age and the Third Temple makes me orthodox in my heart. I believe in the coming of the Messiah and the Resurrection. The bible did not allude to descriptions of the forces of evil existing today in our world. The Devil and his cohorts want to thwart God's plan for the coming of spiritual human beings living in an eternal Messianic age.

A discussion of evil is in order.

18

❧

EVIL
THE DEVIL'S REACH IN SOCIETY

A LMOST FIVE HUNDRED YEARS AGO, THE FIRST LUBAVITCHER REBBE, SCHNEUR ZALMAN, SPOKE ABOUT MANKIND'S CONTINUAL QUEST TO DISCERN GOOD FROM EVIL AND EVIL FROM GOOD. For many centuries the Jews have believed in Gehinnom, which the Christian culture refers to as hell However, hell is not complete unless you assign it a master. The master goes by many names Satan, Devil, Samael, Beelzebub, Lucifer, and so forth. But most people believe that the Devil does not actually exist as a person.

Marcia and I are here to dispel that notion and tell you that the Devil is alive and well and is plotting against God to prevent the Messianic Age. The irony is that God created the Devil although the Devil believes he created himself, and like God he always existed as a separate entity. If you go back to those moments just before the big bang, the Devil would claim that he was the darkness and he was allowing the light of God to exist and not the other way around, where God created the darkness. The Devil is very patient, waiting for centuries for the opportune moment to strike. But he wasn't always a man.

As I wrote in my novel, *Gog and Magog: The Devil's Descendants*, Satan was once a fallen angel who was booted out of Heaven for his attempts at rebellion to overthrow God. He appeared in the Garden of Eden as a serpent almost six thousand years ago to entice Eve to eat of the fruit of the Tree of Knowledge of Good and Evil. From the moment Eve, and then shortly thereafter Adam, took a bite of the apple, or perhaps it was a fig, evil was internalized into man as the evil inclination. From this time forth, the Devil would influence man toward evil in his moral actions, so as the first Chabad Lubavitcher Rebbe, Schneur Zalman, pointed out, evil and good were now embedded inseparably together.

The Garden of Eden is the popular story for God to demonstrate his point, but evil long existed in the modern man, Homo sapiens sapiens, who came about through evolution perhaps fifty thousand plus years before the time of Adam and Eve. You would be correct to ask, why would God create evil? We've seen the consequences of evil in the bloodbaths of human history. Man has been killing man since the days of the caveman, when there were no villages or cities or even farmland and men and women were hunters and gatherers on the savanna. God is not responsible for the killing, but He is responsible for the creation of evil. Man is responsible for all our wars. As Julius Caesar remarked in ancient Rome, *"The fault, dear Brutus, lies not in our stars but in ourselves."* God gave us free will, and we haven't necessarily made the best use of it.

Religious people will argue that our God is a good God, and I wholeheartedly agree with them. God needed for humans to know imperfection in preparation for the perfection of the Messianic Age. The same is true for knowing evil so we will appreciate the goodness of the coming eternal life.

If there were no free will to make the wrong moral decisions, how would we know what the right choices were? If life was just all good as it was in the first Garden of Eden before the eating of the forbidden fruit, we would probably be automatons when we entered the Messianic Age and life would be boring. God created a world of good and evil, but it's a damn interesting world despite the evil imagery and insensitivity of man who can't seem to help himself, despite God's gift of a Divine soul. In the course of life, humans seem to easily be able to darken their Divine souls. It is much easier to blacken your soul than it is to brighten it. Climbing the spiritual ladder takes time and energy. Living in the material world is impulsive because material appealing visuals are being dangled before our eyes.

Marcia always spoke about life in terms of being two sides of the same coin. Good and evil or love and hate or imperfection and perfection are opposites, but are also two sides of the same coin of life. In order to deeply love, at one point in your life it must come out from within you that you can deeply hate. My rage came out in primal therapy when I regressed in time to my mother's womb. Native Americans all believe we have an animal spirit. The animal kills on instinct. We kill fully aware of our actions. We are lower than animals when we do this. Make no mistake, all virtuous people have the capability of hate although they may deny it. God made us this way so that we would know our negative traits when we will be transformed with all positive traits at the End of Days.

I'll put in a plug for my book, *Gog and Magog: The Devil's Descendants,* which is available in paperback and in various eBook formats. Gog and Magog are biblical prophecies, and even though I wrote my book as a fiction, the prophecies are nonfiction and are destined to come true unless God changes them. A brief comment on the Wars of Gog and Magog, which will bring about the cataclysmic destruction of mankind on a scale of devastation that we can't imagine, follows: Let it be said that it is in the interests of the Devil to bring this apocalypse to fruition. I know the Devil exists because he has basically told me his plan in so many words, while he swears profanity at me for loving my God and being a goody two-shoes.

The Hebrew and the Christian Bibles talk about the Wars of Gog and Magog. No one knows for sure who or what Gog and Magog refers to although some have speculated, myself included, as to the significance of these names. My interpretation is that "Gog" is an acronym for "god of the gentiles." The acronym of "Magog" is "male ancestry of god of gentiles." Thus Gog is a leader whose ancestry will give him away. In the bible, God speaks about punishing those who have persecuted the Jews over the centuries. Thirteen prophets, as well as King David in Psalms and the righteous Daniel, vividly foretell of the Final Redemption—Isaiah, Jeremiah, Ezekiel, Hoshea, Joel, Amos, Obadiah, Micah, Habbakuk, Zephania, Chaggai, Zechariah, and Malachi. The wars will be fought as a series of apocalyptic battles, prior to the Redemption. It is not clear what the role of the *Pafkod Rifked*, the Redeemer, is in the Wars of Gog and Magog.

The Messiah, whomever he or she is, may not be directly involved in the actual battles, although he could play an indirect role in urging God to prevent the unimaginable battles of Gog and Magog. I outlined my own scenario in my *Gog and Magog* novel, whereby a good defense is having a good offense as far as combating the Devil is concerned. God needs to come back on the public world stage and go on the offense to eliminate evil and thrust us into the bliss of a new redeemed world.

The armies of Gog and Magog, the descendants of the seventy nations of Noah's three sons, will come toward Jerusalem three times and will reach the city on the third try. A siege will take place, but ultimately G-d will step in to slay the oppressors. Only one-third of the population will survive. Peace will be restored between the ten Northern and two Southern Tribes. The prophet, Micah, will oversee the repentance of the surviving Jews as God gives us a new heart and a new spirit. Redemption means to save from sin or its consequences.

How will the Devil do this? It is very real that individuals have sold their souls to the Devil. They have become total darkness, and they walk around just like you and me, being the nicest people and smiling and loving you. While they are smiling at you and seemingly laughing with you, they are passing their evil energy onto you and destroying your health in the hope that they will accelerate your death. It all sounds as if I'm psychotic and am wildly delusional, promoting one huge conspiracy theory. Movies have been made that address the Devil and conspiracy theories, but I assure you this is real. I can pick these people up by my energy measurements, and Marcia has confirmed and is aware of these evil beings. These evil individuals have infiltrated our families, friends, and our government.

The Devil can transform himself into human form and can then walk around unnoticed in our world. The evil being can be male or female and can be a scientist like myself or a teacher like Marcia or anyone else that suits the Devil's purpose. We call such evil people devil incarnates, for the Devil is recreating himself in human form.

The Devil seems to be paying particular attention to the Pollock family and members of our soul group in the spirit world because all evidence and miracles suggest that we are connected to the Messianic Age.

The Devil also has the ability to send his energy into a foetus, like God does with the Divine soul. When the Devil does do this, he creates the Devil's descendants. If you are neither a Devil incarnate nor a descendant but the Devil owns your soul, then you are one of his demons. It is well known that if you associate with evil, the evil attaches to you. There are people who are associated demons but who have not yet flipped over to demon status. The Devil incarnates can flip an associated demon either voluntarily or involuntarily. The Devil also has a team of witches that can cast spells and curses of negative energy. The curses can be removed readily, but the spells can be more difficult to eliminate by the Yuen technique. There are also entities or ghosts whose bodies died but have not yet returned to the soul world. These ghosts can wreak havoc and be part of the Devil's arsenal. I have had many a sleepless night with bad dreams and visits by ghosts these past few months.

The negative energy from these evil beings can be passed on through touch such as hugs and kisses or a handshake. If a demon has written on a birthday card and you touch the card, you have picked up some negative energy that you must remove. Otherwise the bad energy can damage you. A phone call by a demon can transmit negative energy when you hear his voice on the phone.

I'm saying all of this as if it applies to all of you reading this eerie chapter. It doesn't. Evil energy passage is rare in today's world. Marcia and I are speaking about the evil singularly and deliberately directed against the Pollock family and our soul group and adjacent soul groups. However below we discuss Channeling as a mechanism for the Devil to pass on his evil energy. Marcia and I know who the members of our groups are, but the members of our group don't know because their memories were wiped out at birth. They all came back with Marcia and I in a final incarnation because we are close to the time when we shall be facing the final battle between good and evil. All of our members are old souls who have lived many past lives. Through energy measurements, Marcia and I know who our members are. Our members are also unaware of the evil being acted upon them.

When my daughter Melanie was down last summer, I previously mentioned that I taught her the energy-measuring techniques, and we determined who the people were who had close contact with our soul members and who were either demons, associate demons, witches, or devil incarnates. Melanie sensed evil in a different way than I did, and after perfecting our techniques, we are now pretty much in agreement. Even seemingly close friends and infiltrated seemingly loving family members turn out to be owned by the Devil.

I taught Melanie how to remove negative energy from ourselves and from our soul members and close-by soul group members, after we determined who some of our soul members were. This is done to reemphasize through Yuen's method, as was taught to me by Jules, by feeling a light of white, yellow, gold, or blue running from inside an imagined hollow space within your body from your neck to your genitals and saying "correct neutral" to remove the negative energy. Marcia and I described this in the chapters on Energy Channels and Soul Energy. We can then check by energy measurements to see if the negative energy is removed. The Devil often interferes with me especially when I am trying to remove the bad energy from members of our soul group and close-by soul groups. Sometimes it takes time and repeated attempts to do this, and I have to optimize my soul energy with theirs. I always correct neutral to be completely objective and I also remove interference by the Devil and any of his malevolent intruders. If I don't remove my own thought bias and evil interference, then I won't get the right answers to my questions about good and evil, positive and negative energy, and the soul. Marcia as a free soul in the spirit world doesn't have to do this.

I remove the negative energy from good people by passing the light through them and correcting neutral as I do for myself. I can also send them positive energy from my soul energy, but I can't do this with a demon, who mostly have dark energy within themselves. Marcia, myself, and our daughter Melanie are aware of some of the evil individuals because of my experiences with the soul world under hypnosis regression.

The Devil's plan for Gog and Magog cannot take place without the help of his human cohorts whom the Devil exercises total mind control over. The movie, *The Manchurian Candidate*, fictitiously depicts brainwashing, but the Devil is for real. These evil cohorts have been placed in the highest levels of our society. God has forbidden Marcia and I to speak about whom these individuals are at this time. All we can tell you is that there is a future potential for events to line up in such a way that God will once again take his place on the world stage as He did back in biblical times.

We wanted to emphasize that I wasn't supposed to find out about any of this evil. Marcia's death and my past lives-spirit world regressions seeking her out were destined to lead me in this direction. Sometimes it seems that you waste a lot of precious time in your life achieving or failing to achieve goals that never lead you anywhere. The truth is that everything you do is important and is a preparation for your spiritual mission. I never could have written this book if I wasn't a scientist, had an upbringing that forced me into primal therapy, and I was blessed to experience God's miracles. Nor could I have hoped to achieve what I have achieved without Marcia. She has been my rock and my soul mate throughout the ages. Energy measurements on my part and Marcia's Divine soul indicate that we were soul mates 54,500 years ago. A long time ago!

Perhaps we were the first *Homo sapiens sapiens*. If that were true, we would have received a part of God's energy particles even thousands of years ago as Jerry and Marcia Pollock. This implies then that we would be in our second go-round as ourselves. All the lives that followed us thousands of years ago would then have tiny pieces of our soul, like Moses and his sister Miriam, or George and Martha Washington. Let us be clear once again, we are not our past lives because we receive only 1 percent of their energy, but these past lives certainly are an integral part of who we are.

The Devil's reach is far and broad in our society. During the last thirty to forty years, there has been a renaissance in people communicating with spirits via Channeling which is an ancient practice. We suggest that Channeling is owned and is a creation of the Devil.

Channeling is very different than the direct thought-energy telepathy between Marcia and myself, or a legitimate psychic who acts as a medium for making a connection between those in the physical world love ones in the spirit world. In Channeling there is no direct communication between you and various spiritists or divinists like spirit guides, masters, angels like Archangel Michael, Jesus, even non-physical entities and pets.

What happens is these spiritists send you messages from the supposed spiritual world and you act as a "Channel" to receive them. The message is not given in words but whether you write them down or just listen or speak to them, you must decipher these messages as if you are a translator of a foreign language. Some Channelers have described these messages as a flow of non-verbal communication in the form of an intuitive mixture of thoughts, words, emotions and sensations, jumbled up all together, that the Channeler's mind converts and puts into words.

You can receive these messages in a totally or partially conscious state where you think that the entity does not enter your body or in a light or heavy unconscious trance. In the latter instance, your host energy moves aside allowing the spiritist entity full access to your body and you lose full awareness of what transpires. You may even choose to hang out with your passed love one while the Channeling is occurring. The best Channelers suggest that the messages being sent may not be accurate but when they are, they provide a clarity of spiritual thought that cannot otherwise be achieved.

I have helped others connect to their love ones in the spirit world and recently I helped a woman legitimately contact her father via direct thought-energy telepathy like Marcia and I communicate. We did this on the phone from Florida to California. However immediately following the telepathic communication, this woman also unknowingly Channelled where she felt a sensation and then expressed this feeling in words. When I checked my energy levels I discovered that the Channeling was not through her father but by some evil cohort which turned out to be the Devil.

When I got off the phone, I realized that she wasn't actually interpreting the feeling or thought as is believed by Channelers, but rather the Devil was placing the thought into her brain without speaking to her. She then repeated the Devil's thought to me over the phone. Like Marcia, the woman's father realized that the Devil was interfering. How can the Devil do this and how can he place thoughts into a multitude of Channelers' brains at the same time all over the globe?

As we've stated in this book, the Devil is very powerful because he has both human and angelic attributes. I know that on numerous occasions the Devil can speak exactly as Marcia and I have to check the energy of my Divine soul to be sure whom I am speaking with. The Devil is fully aware of spirituality and could easily pass false messages as his own thoughts using the pretext that these messages are coming from guides, masters and angels. We suggest that people in good faith are being duped by the Devil. Channeling is the Devil's masquerade.

Back thirty-three hundred years ago at the time of the Exodus from Egypt, God warned the Israelites and the generations to follow about Channeling being an abomination of sorcery and witchcraft. God realized that His brand of true spirituality would be muddied and cluttered by the Devil. And God's ancient words have come true because so many people believe that they are contacting the spirit world when they Channel. I personally would recommend and trust only voices of love ones that you know. Even with direct communication with so called "masters," how do you know the master is real and not the Devil? I have a way of checking through energy measurements and I can confirm everything with Marcia's bodiless soul in the spirit world.

The early Hebrew prophets, handpicked by God, were the very first humans to serve as "Channels" for God's words to the Jewish people. The prophets received God's messages while in a trance- or dream-like state and they helped guide the people in their daily living. Only Moses was in a conscious state and wide awake when God came to him and conversed. The later prophets were given a gift to accurately predict the future while they spoke the word of God to the people. Under normal circumstances, the future is a series of probabilities and is impossible to exactly predict.

In biblical times in Egypt, Mesopotamia and the Holy Land with the idol worshipping Canaanite nations, knowledge came not through God but through augury or enchantment, sorcery and evil wizards who used magic and drugs, and witchcraft and charmers who used evil potions accompanied by curses and spells not unlike the voodoo-ism practiced in Haiti. Communication with the dead often involved a human skull.

The messages conveyed by the sorcerers of the past and now with the Devil in Channeling are the same evil. They are the antithesis of what God was conveying through His prophets to follow the Ten Commandments and lead a spiritual life. The Devil is in essence masterfully pulling good people, even believers in God, away from God. In the process, all people who Channel are allowing health-damaging evil energy to enter their bodies through the Devil's thoughts.

Channeling is an illusion created by the Devil. God sternly warned repeatedly against doing this in various passages in the Hebrew Bible. Nowhere in the Bible does it say to trust these spiritual entities over God. The entities become idols that you hero worship when you Channel. The rub is that the Devil is these entities and if you are not careful, he will own your soul. You have been tempted exactly like Eve was tempted by the Devil serpent in the Garden of Eden.

The belief is that during Channeling, your energy is displaced by the energy of the entity that you have allowed to enter your body. This dark energy of the Devil is changing the percentage of your Light soul energy that keeps the body alive. Chronic illness can result and if this is the Devil's energy, you can die when your Light energy drops below 18 percent. People claim to contact angelic entities like Archangel Michael. However, God created the Angels and you need His permission to do so.

God gave the Devil dark energy so the Devil could influence our evil inclination in our left brain and also even become the evil inclination in evil human beings. Through the evil inclination, the Devil can place thoughts simultaneously in all those who Channel. Everyone has free will given by God and you can choose to ignore our message of the dangers of Channeling. The way to God is not through Channeling Jesus in Christianity. It is a direct spiritual belief in Jesus without intermediaries.

What is Resurrection and when will it happen? Biblical scholars have proposed that Resurrection will take place before the Wars of Gog and Magog and before the Third Temple is built. Marcia and I suggest that the Messianic Age will begin when the Devil is dead and the Evil Inclination exists but is subdued in human beings, just as it was in the Garden of Eden for Adam and Eve. God created the Devil so that he would be permanently destroyed at a future time when a Prophet shall arise on Earth. The Devil's days and those of his disciples are numbered.

19

⁓❧⁓

RESURRECTION

I SAVED THIS CHAPTER FOR LAST BECUASE IT REPRESENTS THE ONLY REQUEST THAT I WANT GOD TO FULFILL. I ask Him to restore Marcia to me in physical form as she was before the cancer. I beseech Him from this earthly world now and I asked Him when I was up in the spirit world during my past life regressions. God's answer to me has basically been, "It is not time, Jerry." When I ask Marcia when she will be returning, her best answer is, "Soon, Jerry." However, she may or may not have an accurate time line of what "soon" means, since short-ranged time seems to be irrelevant in the spirit world. I take both of these responses to mean that it will happen. I have to hang in there, although life is lonely and sad without Marcia.

By Resurrection, we are referring to her Divine soul in Heaven coming back into her dead body, decomposed or not, to recreate Marcia as she was in the physical world. This incredible event happens despite the fact that her buried body has likely decayed to a skeleton. Religions like Catholicism have evidence that decomposition of the dead body can be prevented in Incorruptible individuals who are considered Holy. Such Saintly individuals have eradicated the inclination to cause harm to others. My Marcia I know is Incorruptible and it is possible that God has granted her sufficient soul energy which is preventing her body from decaying. This would facilitate her Resurrection. Marcia will be promised youthfulness as she will come back free of all disease, according to the prophecies. This is true for all who will be selected to be resurrected. In the case of cremated bodies, it is not clear if and how the Resurrection will take place because this is not described in the Hebrew Bible. Cremation or traditionally buried individuals in Christianity, or any other religion, may claim a different right of passage at the End of Days. Jews today are also cremated, so it's not clear how the soul comes together, since the skeletal structure of the body no longer exists. Another serious problem is the Holocaust Jews who horrifically died in the gas chambers of Nazi Germany and were then burned to nothingness. This of course will not be a problem for God Whose miracles created Adam from the stardust. God will thus recreate individuals worthy of entering the Messianic Age from their cremated bodies. This will be especially be true for those who cried out to God when their lives were being terminated in the gas chambers of Nazi Germany.

In Daniel 12:13, God tells Daniel, *"And as for you, go to the end; you will rest and arise to your lot at the end of the Days."* God doesn't tell the righteous Daniel that he will enter the Messianic Age. Daniel will be resurrected, as will everyone else, to face his judgment by God. Each one of us will be accountable for our moral choices.

In Daniel 12:13, Daniel gives us insight into what is in store for us. Daniel's visions are both incredible and frightening: *"And at that time, Michael the great prince shall stand up. And it shall be a time of trouble such as never was ever since there was a nation until that time. And at that time, my people shall be delivered, all those who shall be written in the book. And many of them that sleep in the dust of the earth shall awake, some to eternal life and some to reproaches and everlasting abhorrence. And the wise shall be resplendent as the splendor of the firmament and they that turn many to righteousness as the stars forever and ever."*

In Isaiah 26:19, the prophet Isaiah, who was murdered in the Temple by the king's guards, sings: *"The dead shall live. My corpse shall arise awake and sing. You dweller of dust. For a dew of light is thy day. And the earth shall bring forth the shades."*

In the Vision of the Dry Bones, Ezekiel 37:114, a long discourse takes place between God and the prophet Ezekiel. God tells Ezekiel to prophesy so that the dry bones of the dead will be infused with spirit and the dead shall be resurrected and live again. God says that he will bring sinews and flesh to the bones and stretch the skin over the skeletons of the dead. Ezekiel observes this firsthand. God then breathes in a living soul, as He did with Adam, so that the resurrected individual will come to life. Finally, God says, *"I will open your tombs and will bring you up from your graves."*

Because God is eternal, the human soul is Divinely eternal. The body, on the other hand, comes from the seed of one's mother and father. The soul is called the candle of God, by which God examines our internal parts. Our free will determines the moral choices we make in life, which in turn may govern whether we enter the Messianic Age and eternal life. The Resurrection applies not only to those like Marcia in the recent present, but also to all past lives who have passed God's test. Not everyone, as we discussed, has led a sufficiently righteous life; just because one past life has and will be admitted into the Messianic Age, your other past lives will not. According to Newton, souls choose a variety of bodies and not all combinations of souls and bodies in any one life will lead to a good, moral, righteous, and just record.

How God will bring together the body and the soul under all circumstances can't be answered at this time. This will be one of the miracles that God speaks about as being incredible even to Him. It would be nice to imagine a future time of spiritual human beings where the soul and body play an equal role in our lives, so that we can have the best of both worlds and live for the eternity. We hope that you have enjoyed our writings. Our wish for you and us is that we shall be together in the Messianic Age and meet up with Albert Einstein.

I love you, Marcia
I love you, Jerry

FURTHER READING

BUZAN, TONY. *THE POWER OF SPIRITUAL INTELLIGENCE 10 WAYS TO TAP INTO YOUR SPIRITUAL IINTELLIGENCE.* THORSONS: LONDON, 2001.

Chia, Mantak. *Taoist Cosmic Healing: Chi Kung Color Healing Principles For Detoxification and Rejuvenation.* Destiny Books: Rochester, Vermont, 2003

Fehmi, Les. *The Open Focus Brain: Harnessing the Power of Attention to Heal Mind and Body.* Trumpeter Books: Boston, 2007.

Isaacson, Walter. *Einstein: His Life and Universe.* Simon & Schuster: New York, 2008.

Moody, Raymond and Perry, Paul. *Reunions: Visionary Encounters With Departed Loved Ones.* Ivy Books: New York, 1993.

Nelson, Bradley. *The Emotion Code: How to Release Your Trapped Emotions for Abundant Health, Love and Happiness.* Wellness Unmasked Publishing: Mesquite, New York, 2007.

Newton, Michael. *Journey of Souls: Case Studies of Life Between Lives 5th Edition*. Llewellyn Publications: Woodbury, Minnesota, 2003.

Newton, Michael. *Destiny of Souls: New Case Studies of Life Between Lives*. Llewellyn Publications: Woodbury, Minnesota, 2011

Pollock, Jerry. *Divinely Inspired: Spiritual Awakening of a Soul* 2nd Edition. Schechinah Third Temple Publishing: Boynton Beach, Florida, 2009.

Pollock, Jerry. *Messiah Interviews: Belonging to God*. Shechinah Third Temple Publishing: Boynton Beach, Florida, 2009.

Pollock, Jerry. *Gog and Magog: The Devil's Descendants*. Shechinah Third Temple Publishing: Boynton Beach, Florida, 2011.

Ribner, Melinda. *New Age Judaism: Ancient Wisdom for the Modern World*. Simcha Press: Deerfield Beach, Florida, 1999.

Shroder, Tom. *Old Souls: Compelling Evidence From Children Who Remember Past Lives*. Simon & Schuster: New York, 1999.

Stannard, Russell. *Relativity: A Very Short Introduction*. Oxford University Press: Oxford, 2008.

Thie, John. *Touch for Health: A Practical Guide to Natural Health With Acupressure Touch*. DeVorss & Company, Publisher: Camarillo, California, 2010.

Weiss, Brian. *Many Lives, Many Masters: The True Story of a Prominent Psychiatrist, His Young Patient, and the Past-Life Therapy That Changed Both Their Lives.* Simon & Schuster: New York, 1988.

Weiss, Brian. *Only Love is Real: A Story of Soulmates Reunited.* Warner Books: New York, 1996.

Zohar, Danah and Marshall, Ian. *Spiritual Intelligence: The Ultimate Intelligence.* Bloomsbury Publishing: London, 2000.

ABOUT THE AUTHORS

MARCIA POLLOCK IS A GRADUATE OF STONY BROOK
UNIVERSITY ON LONG ISLAND WITH A MAJOR IN
PSYCHOLOGY. She holds a masters degree in special
education teaching from C.W. Post University, also on Long
Island. This is her first book. The uniqueness of her contribution
resides in her co-authoring this book with her husband and soul
mate from the spirit world. Marcia passed on March 18, 2011,
from liver cancer. She retired from the Brentwood School District
in July 2005. Her soul lives on for all eternity.

Jerry Pollock has a bachelor of science and master of science in pharmacy from the University of Toronto. He obtained his PhD in biophysics at the Weizmann Institute of Science in Israel in 1969, and then went on to New York University Medical Center for four years as a postdoctoral fellow and assistant professor of microbiology. Until July 2006 he was professor of oral biology and pathology at Stony Brook University and is now professor emeritus. Dr. Pollock recently completed Marcia's Memoirs, *The Wing of the Butterfly*, for family and friends. He is the author of *Divinely Inspired: Spiritual Awakening of a Soul*; *Messiah Interviews: Belonging to God*; and *Gog and Magog: The Devil's Descendants*.

Marcia and Jerry wrote this book for God, as it is their humble attempt to give back to the Creator for the gifts He continues to bestow upon them.